Praise for
We Are Electric

"In her engaging debut...[Adee] traces efforts to explore—and sometimes exploit—the human 'electrome,' doing full justice to the complex issues surrounding the body's electric forces."
—*New York Times Book Review* (Editor's Choice)

"Sally Adee manages that most difficult feat in science writing: taking a subject you didn't know you cared about and making it genuinely fascinating and exciting. The 'ohmigod-that's-so-cool' moments come thick and fast as she brings the science up to date, investigating today's cutting edge and what the future may hold for bioelectric medicine. It's a vast and hugely exciting area of scientific research, shared with infectious enthusiasm, a real depth of knowledge, a smart and funny turn of phrase. You'll never think of life in the same way again."
—Caroline Williams, author of
Move! The New Science of Body over Mind

"The human body runs on an electricity we barely understand. Unlocking its secrets has the potential to usher in a new age of human health interventions that will revolutionize the way we comprehend and treat our most common maladies. In this fascinating look at this next frontier of scientific discovery, Sally Adee explores the untold history of bioelectricity and sketches its tantalizing, and promising, future."
—Jamie Metzl, author of *Hacking Darwin: Genetic Engineering and the Future of Humanity*

"If you thought genetics was the secret of life, think again: in *We Are Electric*, Sally Adee vividly explores the magic of bioelectricity and how it affects every aspect of our being. A joy to read—I loved this book."

—Joseph Jebelli, author of *In Pursuit of Memory: The Fight Against Alzheimer's*

"As Sally Adee describes with great wit and insight, we are nothing without electricity; it's the stuff of life, and of death. This is such a thrilling, compelling, and energizing book—reading it, I couldn't help picturing the author as Zeus, chucking lightning bolts at me. Such a timely book, too. The future is—I'm sorry, I can't help it—electrifying."

—Rowan Hooper, author of *How to Save the World for Just a Trillion Dollars*

"The electrome may be as important to our understanding of life as the genetic code—yet few of us are aware of these groundbreaking developments. With scintillating storytelling, Sally Adee takes us to into the heart of this scientific revolution and its potential to transform medicine. *We Are Electric* is science writing at its very best—it shimmers with wit and insight. Prepare to be entertained, enlightened, and yes, electrified, by this brilliant book."

—David Robson, author of *The Intelligence Trap* and *The Expectation Effect*

"This book blew my mind. *We Are Electric* is a thrilling read, and Sally Adee explains everything from the intricacies of our electric cells to the potential for new medical treatments—and brain-hacking—with a sparkling clarity."

—Michael Brooks, author of *The Art of More: How Mathematics Created Civilization*

"[In] her excellent first book, *We Are Electric*, about the newly discovered world of the body's so-called electrome...Sally Adee has written an absorbing and fast-paced account of a field of research that could thus herald a whole new era of paradigm-shifting medicine. Moreover, she has done so without apparently drinking the Kool-Aid of today's many bioelectricity boosters....Adee has performed a sterling service in persuading us to contemplate the benefits and possible implications of what seems our inevitable electric future."

—Simon Winchester, *New York Times*

"Fascinating stuff." —Terry Gross, "Fresh Air" (NPR)

"The research [Adee] describes is certainly remarkable, and her enthusiasm for bioelectricity's enormous potential makes *We Are Electric* a lively read." —*Wall Street Journal*

"Adee wades through piles of up-to-the-minute research about the human bioelectrical system and how it can be manipulated, intentionally or not; if nothing else, it will make you marvel anew at your body as an infinitely intricate machine."

—Gregory Cowles, *New York Times Book Review*

"Adee entertainingly introduces the electrome, encompassing all the different roles that electricity plays in living creatures."

—*Financial Times* (Best Summer Books of 2023)

"Adee gives an entertaining account.... Adee's enthusiasm is infectious, and she conveys well the jaw-dropping scale and complexity of this newly discovered 'electrome.' This 'bioelectrical revolution' is more than medicine." —*Times of London*

"In her debut book, [Sally Adee] paints a riveting (and often humorous) picture of two hundred years of research on the bioelectricity coursing through our bodies, from debates over twitching frogs' legs to devices developed to give sensation back to people with traumatic nerve injuries." —*Scientific American*

"One thing readers might not expect from a book that illustrates the intricacies of ion channels: it's surprisingly funny.... energy thrums through the book, charging her storytelling like a staticky balloon. Adee is especially electrifying in a chapter about spinal nerve regeneration and why initial experiments juddered to a halt.... Such implants bring many challenges—like how to marry electronics to living tissue—but Adee's book leaves readers with a sense of excitement." —ScienceNews.org

"A stylish recounting of the story of bioelectricity, its dramatic history, and thrilling possibilities.... Adee is a reassuring guide through this complex and controversial subject. Her technical explanations are exemplary, rendering biological processes comprehensible to those almost entirely uninitiated with the life sciences.... It is astonishing that *We Are Electric* is the first popular science book on this subject; it taps into the magic of science that we are only on the brink of understanding. A book which does its fascinating subject justice: elegant, exciting, and expertly written."
—The Institution of Engineering and Technology, *E&T Magazine*

"The book is packed with...fantastic stuff."
—*IEEE Spectrum* podcast

We Are Electric

Inside the 200-Year Hunt for Our Body's Bioelectric Code, and What the Future Holds

SALLY ADEE

New York

For Ann

Copyright © 2024, 2023 by Sally Adee

Cover design by Terri Sirma
Cover image © Login/Shutterstock
Cover copyright © 2023 by Hachette Book Group, Inc.
frog © Qbertlegion/Shutterstock

Hachette Book Group supports the right to free expression and the value of copyright. The purpose of copyright is to encourage writers and artists to produce the creative works that enrich our culture.

The scanning, uploading, and distribution of this book without permission is a theft of the author's intellectual property. If you would like permission to use material from the book (other than for review purposes), please contact permissions@hbgusa.com. Thank you for your support of the author's rights.

Hachette Books
Hachette Book Group
1290 Avenue of the Americas
New York, NY 10104
HachetteBooks.com
Twitter.com/HachetteBooks
Instagram.com/HachetteBooks

First Trade Paperback Edition: June 2024

Published by Hachette Books, an imprint of Perseus Books, LLC, a subsidiary of Hachette Book Group, Inc. The Hachette Books name and logo is a trademark of the Hachette Book Group.

The Hachette Speakers Bureau provides a wide range of authors for speaking events. To find out more, go to hachettespeakersbureau.com or email HachetteSpeakers@hbgusa.com.

Books by Hachette Books may be purchased in bulk for business, educational, or promotional use. For information, please contact your local bookseller or Hachette Book Group Special Markets Department at: special.markets@hbgusa.com.

The publisher is not responsible for websites (or their content) that are not owned by the publisher.

Library of Congress Control Number: 2022951646

ISBNs: 978-0-306-82663-4 (trade paperback), 978-0-306-82662-7 (hardcover), 978-0-306-82664-1 (ebook)

Printed in the United States of America

LSC-C

Printing 1, 2024

CONTENTS

Introduction   1

Part 1—Bioelectricity in the Beginning
1. Artificial vs. Animal: Galvani, Volta, and the battle for electricity   17
2. Spectacular pseudoscience: The fall and rise of bioelectricity   47

Part 2—Bioelectricity and the Electrome
3. The electrome and the bioelectric code: How to understand our body's electrical language   71

Part 3—Bioelectricity in the Brain and Body
4. Electrifying the heart: How we found useful patterns in our electrical signals   99
5. Artificial memories and sensory implants: The hunt for the neural code   110
6. The healing spark: The mystery of spinal regeneration   150

Part 4—Bioelectricity in Birth and Death
7. In the beginning: The electricity that builds and rebuilds you   187

8   At the end: The electricity that   214
    breaks you back down

Part 5—Bioelectricity in the Future
9   Swapping silicon for squids: 243
    Putting the bio into bioelectronics
10 Electrifying ourselves better: New   266
    brains and bodies through electrochemistry
    Afterword: Gut feelings: The stomach in your brain 296

Acknowledgments   311
Notes   315
Index   353

INTRODUCTION

I was back at the checkpoint. The traffic moved as normal. Bored-looking soldiers waved through civilians on foot, dusty cars, and rickety trucks full of livestock and produce.

Then the Humvee in front of the gate blew up.

Out of the eye-searing blast, I made out the figure of a man running at me, full-speed. He was wearing an explosive vest. I shot him.

A flash of movement to my left revealed a sniper who had just begun to raise his gun. I got him too.

Now a mass of people—maybe seven?—breached the checkpoint, all of them with machine guns. I scanned the group to determine who was closest, who I needed to take out first.

Three more men darted across the roof of a low building that overlooked the checkpoint. I saw them anyway. *Bang-bang-bang.*

There weren't any more after that, only the quiet whistle of the desert wind. Still I waited, calm and alert, scanning the horizon.

The lights came up and the tech walked in.

"What's wrong?" I asked.

"Nothing," the tech said, surprised. "You're done."

"What do you mean, done?" I was disappointed. I couldn't have been inside the simulation for more than three minutes. "Can I keep going?"

"No, it's over."

"How many did I get?" I asked, as I surrendered my rifle and headgear, cutting off the flow of electricity that had been coursing through my brain.

She shrugged. "All of them."

I was in a gray office park in southern California, nowhere near any checkpoint in any conflict. In my hands was an M4 close-combat rifle modified to fire CO_2 cartridges, and while those can pack a bit of a kick, they don't do any damage. The people I was firing at were not real; they had been dreamed up by the programmers of a wall-sized army training simulation.

What was real was the electrical stimulation device on my head. I had signed up to have a few milliamps from a 9-volt battery sent through my skull to test if it would make me a better shot. The scientists' hypothesis was that the electrical current would recalibrate a different kind of electricity in my brain: the naturally occurring bioelectric signals that the nervous system relies on to communicate. By overpowering these delicate natural streams with an artificial shock to the executive part of my brain, they hoped to wrench my mind into a state of alertness and concentration—enough to turn this desk-slumping journalist into a battle-ready assassin.

Back in 2011, I was a writer and editor for *New Scientist*. It was a dream job for which I had recently moved across the ocean. Before that, I reported on microchips and neurotech for a US-based engineering magazine called *IEEE Spectrum*, an inevitable position for someone with my childhood. My dad is a former radio engineer who filled the basement of our family home with intriguing contraptions—circuit boards, candy-coated diodes, and resistors—and a fairly comprehensive mid-twentieth-century back catalog of a science fiction magazine called *Analog*. Part of the reason I became a science writer was to watch the ideas from those old sci-fi stories undergo metamorphosis into real science.

INTRODUCTION

That would also explain why I was obsessed from the moment I first caught wind of this mind-boggling military brain-stimulation experiment. I had seen this technique—known as transcranial direct current stimulation (tDCS)—bubbling around the science press for a few years. Among other intriguing results, it seemed to improve everything from treatment-resistant depression to poor math skills. This flow of electricity, according to the scientists who wired me up, might alter the strength of connections between the neurons in my brain, making them more likely to fire in concert. That natural synchronization is the basis of all learning, and speeding it up with an electrical field would theoretically accelerate the rate at which I could learn a new skill (in this case, turning me into James Bond).

When I caught my first glimpse of this strange new use for electricity in 2009, it was the stuff of obscure medical trials and secret military projects. Today, the notion of wearing an electrical stimulator on your head isn't as foreign as it seemed back then; it's certainly the kind of thing you can imagine someone in Silicon Valley doing for a little extra mental edge, alongside intermittent fasting or microdosing psilocybin.

But it's not just about boosting your brainpower with a volt jolt—there are many other ways electricity is being used to treat the ailments of body and mind. Take deep brain stimulation, a treatment of last resort for Parkinson's disease, in which two electrodes the size and shape of uncooked spaghetti are slid into the deepest parts of your brain to quiet the disease's destructive symptoms. In the wake of its fantastic success, scientists are testing the treatment on other ailments, including epilepsy, anxiety, obsessive-compulsive disorder, and obesity. Then there's the rise of "electroceuticals": these rice-grain-sized electrical implants, clamped around nerves in the body, supposedly interrupt their signals, and in rat and pig trials, appeared to reverse diabetes, hypertension, and asthma. In 2016, outstanding early results in human trials—in which they seemed to reverse rheumatoid arthritis—convinced Google's parent company, Alphabet, to team up with a pharmaceutical multinational on a £540 million venture to tap into your body's electrical signals, to try to treat diseases like Crohn's and diabetes.[1]

So when I saw the opportunity to be a guinea pig in a US Defense Department project, of course I jumped at it, and I wasn't disappointed: my own experience with tDCS was transformative. Getting my neurons slapped around by an electric field instantly sharpened my ability to focus, and by the transitive property, my sharpshooting skills. It also felt incredible—like someone had finally hit the off switch on all the distracting negative self-talk that had, until that moment, been the main provider of my mind's elevator music. I was a convert, and I wanted to preach the positive power of electricity to anyone who would listen.

When my story detailing the experience was published in *New Scientist*, it went viral. The timing was perfect: in the early 2010s, Silicon Valley magical thinking was ascendant and everyone aspired to becoming a Soylent-drinking productivity goblin. Transhumanists were desperate for new ways to upgrade their sad meat bodies. Electricity was now poised to join the suite of tools that could help people override their fundamental human limitations. The article became a fixture on "DIY tDCS" forums where amateur neuroengineers traded circuit designs and equipment specs that would let them overclock their brains in their basements. Media coverage saw promise and peril: the producers of science podcast *Radiolab* were intrigued by tDCS's ability to engineer artificial zen. The writer and anthropologist Yuval Noah Harari put me into his book *Homo Deus* as a cautionary tale, a dire warning of humans trying to engineer themselves into gods. South Korean documentary filmmakers wanted me to speculate on whether neurostimulation would transform the human condition. One interviewer called me the Avon Lady of tDCS.

Not that I was the first journalist to look into the promise of manipulating the body's natural electricity in this way. Since the early 2000s, thousands of studies—many carried out at prestigious universities like Oxford, Harvard, and Charité—have pointed to tDCS as a way to improve the mind. A little bit of electricity enhanced memory, mathematical skills, attention, focus, and creativity—it had even shown promise for post-traumatic stress disorder and depression. The data and the headlines had been accumulating for years, but my gonzo experience took it out of dry clinical stuff and into the "it happened to me!" category. Seeing dollar signs in the combination of intriguing lab results and growing public interest, enterprising start-ups quickly began to hawk their own commercial versions of the brain-enhancing headgear I had road-tested. These cute wearables, which would set you back a few hundred dollars, had little in common with the £10,000 gear in the Defense Department's suitcase. Nonetheless, they were soon adopted by people looking for any bit of extra mental edge, including high-level athletes. Before every game, the Golden State Warriors—a team so unbeatable that it has been accused of "ruining basketball"—wore them in practice sessions to zap their brains into the zone.[2] The US Olympic ski team used headsets in training drills, raising accusations of "brain doping."[3]

And then came the inevitable backlash. Skeptics started to wonder if this was all a bit too good to be true. Curing depression? Better concentration? Longer memory? Improved algebra? Soon a wave of studies began to debunk the previous glut of hopeful findings: to prove that the currents involved in tDCS could have no possible effects on neurons, one group electrically stimulated a cadaver and concluded that it was pseudoscientific bullshit; another looked at all the effects across hundreds of tDCS studies—a meta-analysis—and concluded that if you averaged out all the effects, you'd end up with nothing.

They had history on their side. The skeptics pointed to 200 years of electro-foolery, in which quacks claimed that their various electrical belts, rings, baths, and other contraptions could cure everything from perennial ailments like constipation and cancer to complaints of a more distinctly Victorian flavor, like the loss of "male vigor" and excessive masturbation. To the critics, here was proof that the people hawking the benefits of electrical brain stimulation today had no more science behind them than the charlatans selling electric penis belts in the 1870s.

A consensus emerged that tDCS was, if not outright quackery, certainly in the same postcode. Were they right? Had I become the latest victim of the placebo effect? Had I fallen for reheated, 200-year-old snake oil polished to a silicon shine?

I had started to wonder about this myself. Still mesmerized by the glow of that first tDCS nirvana, I had quickly set out to sample the cranial delights offered at other labs. I found that the experimental psychology department at Oxford was investigating tDCS's potential role in boosting mathematical ability. As this was not, uh, a strong suit, it would be the perfect way to check my potential placebo bias—a repeat experiment to see just how good electrical stimulation was.

I swanned into that place expecting virtuosity. I envisioned my hand casually dancing across a blackboard, populating it with the caliber of equations you see in *Good Will Hunting* and *A Beautiful Mind*. I was excited. But when I left the lab several grueling hours later, the closest thing to a light-bulb moment was the glow of my red face, humiliated by several hours of what had essentially been a public and very poorly executed exam. While wearing a silly electrode cap. I had failed to unlock my inner math genius. Maybe it really all was bollocks.

INTRODUCTION

But if it *was* quackery, why did it still seem to work across such a wide swathe of ailments? Surely all those doctors couldn't be wrong? At the time I was seeing medical electricity research everywhere—and not just the comparatively harmless little tDCS jolts. Invasive stimulators implanted in the spine were helping paralyzed people walk again; implanted in the brain, they were helping people with untreatable depression get out of bed; implanted in the vagus nerve they were curing rheumatoid arthritis. What *was* it about electricity? What mechanism could it be using to heal the body? I couldn't get the question out of my system: what was the relationship between electricity and biology?

If this technology worked, I had no idea how. So I decided to work it out. Once I fell properly down this rabbit hole, it took me a decade to climb back out. I've spent the last ten years of my life being electrified by these questions and their answers—and now I want to pass that jolt onto you.

We Are Electric is about the natural electricity that surges through all our bodies, and the truly head-spinning ways the world will change if we learn how to manipulate it. Over the next few hundred pages I will teach you about the substance that courses through all living things, underpinning their every move and intention. This natural electrical current predates nervous systems and even humanity itself; it was coursing through the bodies of our ancestors long before the first fishy mutants even squelched onto dry land. It is the most ancient thing about us. It is among the most ancient things about life itself.

My brief foray into professional marksmanship is just one example of the promise and peril of harnessing our body's natural electricity. We are fundamentally electrical creatures, but the full extent of our electrification would shock you. It is hard to overstate how wholly and utterly your every movement, perception, and thought are controlled by electrical signals. This is not the electricity that comes from a battery or the kind that turns on the lights and powers the dishwasher. That kind of electricity is made of electrons, which are negatively charged particles flowing in a current.

The human body runs on a very different version: "bioelectricity." Instead of electrons, these currents are created by the movements of mostly positively charged ions like potassium, sodium, and calcium. This is how all signals travel within the brain and between it and every organ in the body via the nervous system, enabling perception, motion, and cognition. It's fundamental to our ability to think and talk and walk and why our knee hurts after a fall, and why the scraped skin heals. It's what makes gummy bears taste sour, why we can pick up a glass of water to wash away the taste, and how we know we were thirsty in the first place.

The stuff that comes out of your wall socket is created by a power plant. For the stuff in your body, the power plant is you. Every one of the 40 trillion cells in your body is its own little battery with its own little voltage: when it's at rest, the inside of a cell is (on average) around 70 millivolts more negatively charged than the extracellular soup outside. To keep it that way, the cell is constantly shuttling ions in and out of the membrane that surrounds it, always striving to maintain that -70mv. All of this may sound very petite to you, somewhat beneath your notice. And, yes, at the scale of our lives, a difference of 70 millivolts is insignificant; it's about one thousandth of the amount of electricity required to power a hearing aid. But from a neuron's perspective, it is anything but. When a nerve impulse comes roaring down a nerve fiber, channels open in the neuron and millions of ions get instantly sucked through them into and out of the extracellular space, taking all their charge with them. The electrical field generated by this mass migration of charge

INTRODUCTION

works out to about a million volts per meter, which at that scale would feel like passing *an entire bolt of lightning* from one of your outstretched hands to the other. That's what it feels like to be every neuron in your body, every moment of your life.

Biologists have known for a long time that these kinds of bioelectrical signals are responsible for all communication between the brain and the nervous system: you can think of them as the telephone wires that help the brain's command center communicate with your muscles to operate your limbs.

But bioelectricity isn't confined to our brains. Over the past couple of decades it has become clear that these signals are pressed into service by every cell in your body, not just those that govern your perception and motion.

Each of your skin cells has its own voltage, which it combines with neighboring cells to generate an electrical field. You can even measure your skin's electricity with a voltmeter: just stretch a piece of skin and connect it to electrodes and the "skin battery" will light up an electrometer. You could power the same bulb with a prostate battery. Or a breast battery. When that field is disrupted by injury, you can feel it. That tingling when you bite your tongue or the inside of your cheek? It's the wound current, calling to the surrounding tissue to send help.

Similarly, the cells in your bones are electric. Your teeth are electric. Your organs are electric—and so is the coat of epithelial tissue that hugs each one. Blood cells too. Every single one is a microscopic power plant generating a tiny voltage to communicate within and among themselves.

We used to think those non-nervous-system cells mainly used bioelectric signaling for trivial janitorial and maintenance tasks— for example, waste disposal and energy management. But new research is making it increasingly clear that they do far more. You and I are so much more electric than is commonly acknowledged.

Recently it has been discovered that electrical signals also send out beacons as we grow in the womb to guide us into the eventual shape we will take—two arms, two legs, two ears, a nose. When this signal is interrupted in utero, things go terribly wrong, so scientists are now working out ways to prevent physiological birth defects by retuning our electrics. And what we can do for birth, we can also do for death: cancer cells have their own unusual voltage, and recent evidence indicates that they use electric signaling to communicate about their host environment. Disrupting these signals could keep cancer cells from metastasizing.

Nor is this natural electricity confined to us animals—the same signals have been detected in everything from algae to *E. coli*. Plants use them to send messages across far-flung parts of themselves, warning of predators and turning on defenses.[4] Fungi use them to communicate when their delicate tendrils have sussed out good sources of food. Bacteria use them to make decisions about when to grow their communities into antibiotic-resistant strongholds. Even organisms we haven't quite figured out how to taxonomically classify—we shove them into an all-purpose box labeled "protists"—use these electrical communication signals.

I tell you all this to underscore that "bioelectricity" is no mere metaphor, no elegant stretching of a humdrum biochemical truth. You and I are literally electric. The basis of all life is electric. When our cellular battery runs out, we all die.

So what if we learned how to control the switch?

If you still can't quite get your head around this (or remain suspicious of my enthusiastic contentions), you are not alone. The entire history of bioelectricity has been marked—and in some ways defined—by the skepticism leveled at the researchers from both the physics and biology establishment.

INTRODUCTION

History is filled with tales of the uphill battles biologists faced when trying to suggest that biological phenomena have electrical underpinnings. Today, looking at EEG readouts of the brain's activity is commonplace, but you might not be aware of the ridicule its inventor, Hans Berger, endured—or that he ended up dying by suicide in 1941 before he could see how his device changed the world. Even the most quotidian electrical functions of electricity in the body were only accepted after a knock-down, drag-out fight; in the 1960s, Peter Mitchell spent ten years and a great deal of his own money to build his own lab to convince the scientific establishment that electricity is central to the way a cell generates energy. (He was one of the few to live long enough to see his ideas achieve acclaim, receiving the Nobel Prize for Chemistry in 1978.)

Maybe all this skepticism can be traced to the contentious battle that attended bioelectricity's origin story. Luigi Galvani's discovery in the late eighteenth century that electricity is what lets us move our muscles is perhaps the original electro-controversy: you may have heard of his experiments zapping frogs, but you might not know that doubts about his findings started a scientific war that divided all of Europe. This origin story of bioelectricity profoundly shaped the way subsequent generations of scientists would approach the topic, not least by shaping the structure of science itself. As a result, the scientific knowledge of the electrical underpinnings of life is now scattered across a wide range of disciplines, many of which think the others are peddling poppycock.

Even today, many *bio*logists probably don't know the whole story of *bio*electricity. In 1995, when Mustafa Djamgoz, a cancer researcher at Imperial College London, first proposed his theory that electrical signals were involved in cancer, his colleagues openly dismissed his ideas. Even now, with research awards piling up around him, Djamgoz frequently finds himself re-explaining his research—and needing to start from scratch because sometimes the same concept that elicits a "well, obviously" from one researcher sounds like science fiction to another.

This reflects a set of calcified notions embedded in the framework of science: biologists stick to biology, leaving the study of electricity to the physicists and engineers. They just don't speak the same language. "If you major in biology, you get maybe half a semester of physics, if that," says another cancer biophysicist, Richard Nuccitelli. "You don't even touch electrical engineering." And forget about computer science. This might seem like an obvious and unproblematic division of labor, but it means aspiring PhD students in physics are taught about Tesla and his alternating currents, but not about the bioelectricity running through their own bodies—and biology students get neither. This tacit assumption that each field should "stay in their lane" has been putting limitations both on biology and scientific advancement for decades. What we need is a new framework to bring the body's different electrical parameters under one roof and study them coherently, together.

Call it the electrome.

The identification of the genome and microbiome proved crucial steps to understanding the full complexity of biology, but some scientists think it's now time to plot the outlines of the "electrome": the electrical dimensions and properties of cells, the tissues they collaborate to form, and the electrical forces that are turning out to be involved in every aspect of life. Just as decoding the genome led us to the rules by which information like eye color is encoded in our DNA, bioelectricity researchers predict that decoding the electrome will help us to decipher our body's multilayered communications systems and give us a way to control them.

INTRODUCTION

Over the past ten to fifteen years, experiments have suggested that not only can we decrypt this code—we may even be able to learn to write it ourselves. Researchers are looking for precise ways to flip the circuits inside our cells that are responsible for everything from healing to regeneration to memory. When healthy cells turn cancerous, for example, their electrical signaling changes radically. But restoring these electrical signatures to normal has kicked tumor cells into reverse, to become healthy once again. Other experiments indicate that certain patterns of electrical activity in the brain form specific sensory experiences, and that these may be recorded and overwritten. This could help create advanced prosthetics that a person can feel as fully as they can feel the skin they were born with. If cells really do carry different kinds of messages in their bioelectrical communications, then cracking their bioelectric code might solve some problems that have remained unmoved by all the genetic and chemical interventions we've thrown at them. It would be like opening the electrical box and being able to rewire our systems as we like.

If it were to become possible to manipulate bioelectricity at its source, the consequences would be staggering. Could we interpret these codes well enough to fix our biology when it breaks? Some bioelectricity researchers go as far as to say learning the rules of this software could render our bodies and minds as programmable as hardware. They've mooted all kinds of possibilities: editing people's electrical code to increase intelligence; to reprogram troublesome personalities; to regrow amputated limbs; or to remap the body's blueprint altogether. If we are truly electric, then we should all be programmable at the level of the cell.

But what will happen when we begin to use our knowledge of the electrome to get better grades instead of cure cancer? The gene-editing technology CRISPR ushered in a flurry of worries about designer babies, and our ability to edit the bioelectric code will be much the same. A simple tweak to the electrome caused functional eyes to grow on a frog's butt, and in another prompted a worm to grow two heads.[5] There is a clear relationship between our electrome and the shape our bodies take—from frogs to worms to humans—so we need to do much more research before someone grows themselves a third eye for social media clout.

Bioelectricity research could all too easily be misappropriated by that vague but undeniable urge to see humans as occupiers of inferior meat bodies that could only be improved by the addition and substitution of hardware and software. You know, the idea that someday we'll pop our consciousness into the unblemished silicon Heavens of the Cloud. So what limitations should we place on upgrading or altering humans? Who will govern the rules on remapping the body's electrical wiring? What happens when a brain implant becomes a precondition for employment?

This book will help you understand bioelectricity, both in the brain and nervous system where it has traditionally been understood to work, and in the broader and more unexpected contexts now being found. It will illuminate why we have been applying artificial electricity to tease out how the biological kind works. You'll meet the researchers moving beyond artificial electrical stimulation to build new implants that can talk to our bodies in their own language—from robots made out of frog cells to new electronic implants made from shrimp chitin. If we are going to try to manipulate the human body, the least we can do is manipulate it on its own terms—terms that were honed by millions of years of evolution, and not with headgear we invented. We have arrived at a new stage of bioelectricity. "With bioelectricity, we are now at the point where astronomy was when Galileo first used the telescope," says Djamgoz, one of the cancer researchers gazing into the unknown. If the nineteenth century was referred to as the "electric century," the twenty-first century could go down in history as the bioelectric century.

PART 1

Bioelectricity in the Beginning

"Consider: the hero endures; even his downfall merely foretells his eventual rebirth."
Rainer Maria Rilke, "First Elegy"

Normally, it's difficult to muddle together a coherent story out of all the complex mixtures of culture and history that go into making something the way it is today. But in the case of the confusion about bioelectricity, there's an identifiable chain of causality: a savage battle that helped split science into the constituent disciplines we see today, pitting biologists and physicists against one another in a death match that ultimately determined who gained custody of electricity. Biology lost, physics won, and the consequences would ripple through the next 200 years of science. This original schism profoundly shaped the way subsequent generations of scientists approached the idea of electricity in biology.

CHAPTER 1

Artificial vs. Animal:
Galvani, Volta, and the battle for electricity

Alessandro Volta was astonished. In his hands he held an early print of a manuscript whose author claimed to have solved an ancient mystery: what is the substance that courses through all living things, underpinning their every move and intention?

The answer: electricity.

Volta—a compactly built striver, prone to high, flamboyant collars, whose encroachment of thick black hair was engaged in furious battle with his forehead—felt himself uniquely qualified to evaluate this author's claims. A little over a decade earlier, in 1779, he had been elevated to Chair of Experimental Physics at the University of Pavia, after devising a new tool that dispensed a ready supply of static shocks. It had seen wide adoption by other scientists (and foreshadowed the device that would later cement his name in history), but their smattering of weak plaudits wasn't enough. Volta wanted more acclaim. He deserved more acclaim. He had climbed the ranks, toured the most important scientific centers, and built himself an influential social network of patrons comprised not just of scientists, but politicians and others in the top strata of Italian society. He was on the cusp of establishing himself as one of the global authorities on the controversial, wildly glamorous, and brand-new study of the mysterious phenomenon of electricity.

Electricity was—is—a force of nature, whose mysteries were then just beginning to yield to scientific inquiry. No one understood very much at all about this invisible fluid. It shocked people, it sometimes killed them from the sky, and it was still very much a matter of debate whether it was the same stuff electric fish used to stun their prey. Electricity was also just then in the process of emerging out of the realm of party tricks and ludicrous speculation (a standard-issue claim was that men with strong electricity could produce sparks during sexual intercourse). The first rudimentary tools had only recently been developed to contain this wild stuff for serious scientific investigation and experimentation. Their inventors were eighteenth-century science's version of rock stars. Volta was among them, and had acquired a reputation as a rising star among the scientists who were decoding the mysteries of electricity into empirical truths. Some of his fellow physicists were even starting to refer to him as the "Newton of electricity."[1]

But now this author, the anatomist Luigi Galvani, claimed to have found a biological variant.

Galvani was an uptight bumpkin from an Italian state that had only recently begun to acquire the equipment to bring it up to speed with the current century. A pious obstetrician whose manuscript was full of unsophisticated terminology. *This* person claimed superior knowledge of the stuff that had confounded the smartest men in philosophy and science?

You can sense in the manuscript that Galvani knew the magnitude of the claim he was staking. "We could never suppose that fortune were to be so friendly to me, such as to allow us to be perhaps the first in handling, as it were, the electricity concealed in nerves," he wrote in the preface, with a trepidation that bordered on foreshadowing.[2] Indeed, the claim would eventually rain down ruin.

ARTIFICIAL vs. ANIMAL

How could Galvani's claim—that the body is animated by a kind of electricity—have been so controversial? To understand why Volta became so incensed, we need to understand how far biology lagged behind physics in the late 1700s.

The scientific revolution in Europe had upended scientists' understanding of the physical world by tearing down received wisdom and replacing it with testable laws and predictive equations. Copernicus and Galileo plucked our planet from the center of creation and set it into an unremarkable corner of the cosmos. Kepler discovered the laws governing how the planets moved around the newly central sun. And from these, Newton deduced the law of gravity, and extrapolated how things fall down to Earth.

Biology, on the other hand, discovered few new insights of this magnitude.[3] A promising century ended in an impasse for the study of living things. Microscopes allowed physiologists to examine the minutiae of bacteria, blood cells, and yeast. Anatomists developed detailed maps of the nerves that infiltrated every extremity of the body. It was even understood that these nerves were closely involved in our ability to move our limbs. But how? In the late 1700s, scientists still knew next to nothing about the mechanism that allowed humans to walk and talk and wiggle their fingers and toes, to feel or scratch an itch. How did the immaterial soul direct the motions of the animal machine? No one had the faintest idea.

To say the seventeenth-century understanding of this phenomenon was stuck in the dark ages would be an understatement. It had become stuck much earlier than that—with Claudius Galen, a brilliant physician and philosopher influential in second-century Rome.[4] He kicked off 1,500 years' worth of philosophical musings about what was flowing through our bodies to make us think and move.

Galen's conjectures were aggregated from centuries of Aristotelian thought and refined with the help of a host of dissected cadavers. Nerves, he concluded, are hollow tubes that send man's will by way of ethereal substances called *pneuma psychikon*—"animal spirits"—to be executed in his limbs and muscles; and that is "animal" not in the zoological sense, but in the sense of *anima*, the Latin translation of *psyche*, the Greek word for vitality. These spirits, Galen proposed, are produced in a complex series of interactions inside the body, starting in the liver, distilling in the heart, reacting with inhaled air, and finally being sent to a staging area in the brain.[5] When motion was required, the brain would function like a hydraulic pump to deploy these animal spirits into the hollow nerves for distribution to all the body's feeling and moving parts. When they flowed thus from brain to muscle, the spirits created contractions there. When they flowed in the opposite direction, they carried sensation.

Apart from increasingly baroque refinements, this dogma remained largely unchallenged for at least the next 1,300 years. Any theoretical advances in the field came to depend not on experimental probes, but on philosophical reasoning. For example, in the mid-1600s, René Descartes—the progenitor of mind–body dualism—conjectured that instead of "fire air" the constitution of animal spirits was probably closer to a liquid, like water driving machinery. Medical scientists didn't fare much better. The Sicilian physiologist and physicist Alfonso Borelli proposed that rather than being watery, animal spirits were in fact a highly reactive alkaline "marrow"—in his parlance, *Succus nerveus*, or nerve juice—that squeezed out of the nerves at the slightest perturbation. When this juice reacted with the blood in the muscle, it would cause the surrounding tissue to boil.

These interpretations all ran into the same problem—with the invention of the microscope at the turn of the seventeenth century, it soon became abundantly clear that nerves could not be hollow. That meant there was no room for animal spirits or nerve juice to be the substances governing our limbs. But while these early microscopes were powerful enough to rule out tubes, they were still too weak to probe nerve structure more precisely. This left a crucial question unanswerable: how could anything be transported through a body without the help of tubes? New theories rushed in to fill the vacuum.

Lack of evidence opened the debate to all comers, from the sublimely credentialed to the sublimely questionable. Isaac Newton suggested that the brain's messages traveled along the nerves by vibration, the way you might pluck a guitar string. At the other end of the spectrum was the conjecture of a spa physician in Bath (these were doctors who took up residence at spas, then at the height of their popularity in England, to prescribe exact drinking and bathing regimes—for a robust fee, of course): David Kinneir claimed in a 1738 tract that, as the animal spirits were carried in the blood, taking the waters at the spa would help to unblock the vessels that carried them.[6]

It's worth noting that before the nineteenth century, science was a lot less fussy about its academic boundaries. There was less of a demand back then on the people who studied the natural world to squeeze themselves into rigid disciplines, largely because these didn't yet exist. All that would come later. In fact, scientists weren't even called scientists. People who studied the natural world referred to themselves as natural philosophers, or sometimes experimental philosophers. The ultimate archetype was Alexander von Humboldt, who traveled the world studying whatever took his fancy. Men like him, and Galvani, were free to investigate whatever piqued their interest, which could (and did) range from bone structure to comparative anatomy to electricity.

Especially poorly defined were the distinctions between the physical and life sciences. Cross-field mobility was the norm. Try to categorize people who studied biology in the eighteenth century and you'd be forced to include everyone from radical theologians to physicists. One thing was clear, though. Medics—who were charged with dispensing practical remedies—did not enjoy high status, owing to an increasing awareness of the gap between their scientific airs and their actual ability to treat the sick.

A new hope

By the 1800s, we knew little more about our bodies than we had a full millennium earlier. Meanwhile, the scientific revolution had taken the understanding of electricity from strength to strength.

Like animal spirits, electrical phenomena had been observed for centuries without generating great insight. The Ancient Greeks, for example, had noticed strange stones that seemed to pull metal to them as if by an invisible force. They had seen that when lightning struck people, it often killed them. Electric eels were known to deliver a fulsome shock to their prey. Then there was amber—the insect-trapping resin that also had a strange tendency to attract bits of dust and fluff, the same way the stones attracted metal. Give the amber a vigorous little rub, you might get a little zap and see a spark. But before the seventeenth century, all these observations had not been compiled into any kind of explanatory framework.

In fact, electricity got its name long before we understood how it was involved in any of the above. The word was coined in 1600 by William Gilbert, who—in keeping with what I mentioned earlier about the disciplines—identified as a physician, physicist, and natural philosopher. He borrowed from the Ancient Greek word *elektron*, for amber, owing to that material's unique ability to reliably elicit the magic spark.

The scientific revolution vastly upgraded the tools to investigate the phenomenon. In 1672, Otto von Guericke invented the first device that made it possible for scientists to generate electricity themselves: an "electrostatic generator" was a glass globe you could rub with silk to accumulate a small amount of electrical charge. Touch it and you'd get a zap. (This, incidentally, is where we get the phrase "static electricity." The globe trapped electricity on its surface so it wouldn't go anywhere—it didn't move. It was in stasis.) Electrostatic generators allowed the accumulated electricity to be dispelled in bigger jolts than amber, and that allowed people for the first time to decide how, when, and where to direct the jolts. More machines followed, some making it easier to charge the generator by having hand cranks, so your arms wouldn't get tired from all that rubbing glass with silk. Bigger glass tubes yielded stronger jolts. The shock they generated was weak, but it was enough to start a century of parlor-game science, from the "kissing Venus"—an electrified woman whose kisses stung gentlemen's lips with a trivial zap—to young boys charged up to attract bits of paper and other flotsam as if by magic.

But all of these generators had the same problem: the very act of touching these sources of accumulated static electricity released it all in one go (which is also what's happened when touching your doorknob zaps you with a sharp spark of pain). There was no way to store up a large quantity of electricity for later use.

About a century after the first electrostatic generator, several scientists separately converged on the idea of a special jar that could siphon the mysterious invisible substance from a generator and store it for later. To avoid the thorny question of paternity, the new invention was dubbed the Leyden jar, an indirect credit to Pieter van Musschenbroek, who did a lot of the early work in this Dutch city. Scientists competed to see who could concentrate the most electricity in their jar, because of course they did, and this had exactly the unfortunate consequences you might expect. When van Musschenbroek stuffed his Leyden jar to capacity, rather like overpacking a suitcase, it promptly exploded on him. The shock was enough to send the temporarily paralyzed physicist to bed for two days.

As people got better at stuffing these increasingly capacious vessels to capacity, Leyden jar demonstrations grew progressively more dramatic, from a crowd of 200 monks connected by lengths of iron wire and shocked by a single Leyden jar, to a practical joke in which a specially designed wine glass was electrified for the amusement of picnic guests (less fun for its unfortunate target).[7] Though high society loved these demonstrations, even they agreed that electricity was at best a novelty, and no one could quite deduce how this circus of wonders might prove useful . . . until the mid-1740s, when a Scottish electric showman called Dr. Spencer sent his apparatus to the Philadelphia residence of a young Benjamin Franklin.[8]

Franklin is often credited with single-handedly turning the carnival of electricity into a science. And while it is a bit more complicated than that, Franklin's famous kite demonstration did begin the unification process that proved that different electrical phenomena—lightning, amber, electrostatic generators—were just different manifestations of the same ethereal substance.

Franklin—famous polymath and politician—was among the vanguard of investigators trying to develop a grand unified theory of electricity that would link "natural electricity" (lightning) to the stuff produced by generators and stuffed into Leyden jars ("artificial electricity"). He attached a key to a long string, suspended by a kite, during a lightning storm. If he could charge a Leyden jar with the proceeds of a lightning storm, his point would be proved. It was a stupendously dangerous experiment, but it worked so well that kids are still forced to read about it in school. Upshot: lightning *was* electricity.

Franklin's experiment was hugely consequential and helped pave the way for a new understanding that was formalized into a branch of science, whose practitioners referred to themselves as electricians.

ARTIFICIAL vs. ANIMAL

(This word carried a rather more glamorous connotation back then—you can think of eighteenth-century electricians as the "rocket scientists" of their day.) What's more, there was now an understanding of electricity as an invisible fluid that could be collected in a jar, cross vast distances, and travel along strings, hollow or not.

What else was electricity? By 1776, people had begun to wonder if this "immaterial fluid" wasn't germane to those animal spirits everyone had been wondering about. That year, the notion got its first bit of supporting evidence when John Walsh experimented with an electric eel.

Walsh was a classic natural philosopher: colonel, MP in the House of Commons, all-round rich person. He moved in the same circles as Franklin, who was just starting to cultivate an obsession with electric fish. After their electrical organs had been described, Franklin became convinced that the shock the creature delivered was another manifestation of the phenomenon of electricity, so he convinced Walsh to "devote his scientific energies" (read: a boatload of his ample fortune) to devising experiments that would prove "fish electricity" was real.[9]

The way to do that was to put an electric fish into a dark room and get it to deliver its jolt—in the hope that doing so would yield a visible spark. That would be the smoking gun. Incredibly, it seems Walsh was able to do it. Several historical accounts by people deep in the audience at his 1776 demonstration reported this convincing evidence that electric eels were, in fact, electric. The *British Evening Post* reported "vivid flashes."

While the experiment didn't provide any direct evidence of a link between "fish electricity" and anything that might be involved in human processes, the idea was out there nonetheless: a form of electricity might be at work in the action of nerves and muscles. If an eel could make a spark, perhaps we could create our own internal sparks.

And that's how electricity found Luigi Galvani.

The man who wanted to know God's secret

Historians don't know very much about Luigi Galvani's family and youth. We do know that he was born in 1737 in the Papal State of Bologna, a wealthy and progressive state of Italy. According to the historian Marco Bresadola, Galvani was born into a merchant family; his father, Domenico, was a goldsmith on his fourth wife (Barbara) and second round of children by the time Luigi entered the world.[10] The Galvanis had enough money to send more than one of their children to obtain a university education, which was not an inconsiderable expense. But having a scholar in the family was a mark of social standing and prestige for the merchant classes, so Domenico trotted his kids off to school.

Luigi was initially opposed to this fate. He was a dreamy child who preferred family life to Bolognese student antics. What he liked most was spending time in conversation with the monks at a monastery near Bologna, who were tasked with counseling the dying in their final hours.[11] Galvani was fascinated by the insights the monks brought back from their time with people at the edge of life and death. There, Galvani also absorbed the values and ideals of the progressive Catholic Enlightenment, including the reigning Pope's theories of "public happiness." Instead of focusing on ritual and splendor, as many of his predecessors had, the progressive Benedict XIV tried to inspire his citizens' devotion by actually improving their lives, which took the form of civil engineering projects like public drainage, but also improvements to the education system, including stocking universities with the latest tools, including electrical ones.[12] He redefined faith as charitable action, not competitive superstition.

This philosophy resonated with the young Galvani, and when he was a teenager, he asked to join the order. However, his family convinced the monks to talk him out of it, eager to divert this obviously gifted child onto a more socially mobile track. So Galvani withdrew his inquiry and instead enrolled at the University of Bologna to study medicine and philosophy. (He also studied chemistry, physics, and surgery.) His father was right about his potential—Galvani would go on to write twenty theses just about the structure, development, and pathology of bones. After obtaining his doctorate, Galvani began to research and lecture in anatomy at the university. Though not a natural extrovert, he was a popular lecturer.[13] He was one of the first professors to enliven his talks with experiments, and his enthusiasm was so infectious and his teachings so accessible that students from the neighboring arts academy often crowded into the room. Galvani was awarded a fast succession of academic positions and honors at the University of Bologna, and soon held a concurrent appointment with the Institute of Sciences of Bologna, one of the first modern experimental institutions in Europe.

But he would never quite lose sight of the road not taken—according to all accounts, he remained a devout Catholic to the end of his life. If he couldn't devote himself to God in the monastery, he at least wanted to do it in the laboratory. He lived his principles as best he could, turning his work into an expression of his devotion. In addition to his post at the university, he became a practicing physician at the local hospital. He gave preferential treatment to people in extreme poverty—especially women. As an obstetrician, Galvani nurtured a deep, abiding obsession with creation. More than anything he wanted to understand the scientific underpinnings of how God had given humans the spark of life.

Galvani was in the ideal place, at the ideal time. Founded in 1088, not only was the University of Bologna the oldest university in Europe, it was also the most progressive and forward-thinking.

For example, the university had recently promoted Laura Bassi, its first female lecturer in experimental physics. Bassi was a prodigy who taught Newtonian physics from her home laboratory and established ties with electricians all over the world, including Benjamin Franklin and Giambattista Beccaria, who were considered the leading electrical theorists of the time.[14] This network ensured the university was at the vanguard of this important new phenomenon. Unlike some of his contemporaries, Galvani was in no way scandalized by women in authority or in the sciences writ large; while no one could attach the anachronistic label of feminist to him, he was impatient with the idea that it was "laughable" to take instruction from women. For example, he was nonchalant about his collaborations with the wax sculptor Anna Morandi, whose exquisite anatomical models he used to teach his anatomy class,[15] even as some colleagues blanched at the idea that a woman might have anything to teach them.[16] Untouched by such prejudices, Galvani attended many of Bassi's lectures, and soon she and her husband, the medicine professor Giuseppe Veratti, became his mentors.

At the height of his influence, Giambattista Beccaria sent them his textbook, in which he—like Franklin—was beginning to outline his own grand unified theory of electricity. Beccaria cautiously explored the idea that perhaps natural electricity could be present in animals, having read John Walsh's explosive new publication detailing the anatomy of electric fish. Bassi and Veratti began to encourage their protégés to zap animals with Leyden jars, offering up Bassi's lab for electrical tests on the hearts, intestines, and nerves of frogs.

In Bassi's lab, Galvani grew increasingly obsessed. He began conflating animal spirits with the electric fluid in his lectures. In one anatomy talk on causes of death, Galvani claimed it was rooted in the extinction of "that most noble electric fluid on which motion, sensation, blood circulation, life itself seemed to depend."[17]

While many scholars were beginning to converge on this kind of interpretation, they tiptoed around the conclusion cautiously, loaded as it was with unscientific associations. The more practical problem was that there was no experimental way to test the hypothesis. Still, Galvani was transfixed by the notion that electricity—the stuff in lightning—might be the same mechanism by which God had given breath to man and all other creatures. He was equally transfixed by the notion that he could be the first to discover this facet of God's beneficence.

So, in 1780, he created a research program on the role of electricity in muscular motion, and then set about building a home laboratory that would allow him to spend more time on these experiments. The lab contained an electrostatic machine, a Leyden jar, and other more recently invented variants on this electrical equipment.

From there, he began to experiment on frogs. Why frogs? Their nerves are easy to locate, their muscle contractions are easy to see, and can continue up to forty-four hours after the frog has been carved into the grisly configuration that Galvani referred to as his "preparation." Graphic illustrations of the amphibian experiments suffuse all of Galvani's publications. One shows a frog with its head and midsection almost entirely missing, save for the exposed gossamer strings of the two crural nerves that still connect its legs to its spine.[18] In others, the frogs are cut in half below the upper limbs, then skinned and disemboweled. Only their legs remain, joined to each other by a nub of spine. In another, Galvani and his scientific collaborators, Giovanni Aldini (his nephew) and Lucia (his wife), stand in his basement lab, surrounded by dozens of these flayed corpses.

This very particular method of preparing his frogs—from which Galvani never deviated—was inspired by Lazzaro Spallanzani, one of the most important naturalists of the time and Galvani's frequent correspondent. Spallanzani's specifications made it extremely easy to distinguish cause and effect. With everything but nerve stripped away, there could be no confusion about what happened when you put electricity into a muscle or nerve.

Galvani started his research with a series of experiments designed to help him understand why electrical current from artificial sources caused muscle contractions. Applying a zap to a muscle obviously caused the muscle to twitch, but by what mechanism? At first he simply repeated previous experiments, touching an electrical contact to various parts of the frog's body. To send the electricity from the generator into the specific parts he wanted to target, he used wires and other metal objects called arcs, connected to the source of external electricity speared into various parts of the frog.

Usually the results fell in line with his expectations . . . until one day they didn't. That day, a frog jumped even though there had been no contact between it and a generator. Galvani had been touching the exposed crural nerve of the frog as it lay on its plate. At the same moment, standing about six feet away, Lucia brought her finger close to the machine, which unexpectedly elicited a spark. The frog twitched. Galvani was shocked. In the absence of the usual connections between generator and frog, he could not conceive of any obvious way the electricity could have been transmitted into the dead animal. How could it have twitched without any external electricity to animate it?

No existing hypothesis offered any satisfying explanation, and from that moment on Galvani was "inflamed," as he would later write in his manuscript.[19] He began obsessively repeating variations on the experiment, using any available source of "artificial" electricity—Leyden jars, electrostatic generators—and moving the frog by turns closer and farther away. The frog jumped every time.

This led Galvani down a few blind alleys. First he thought there was some kind of atmospheric electricity in the lab that built up in the frog and was then released when the frog's leg was touched. In 1786, Galvani decided to set up a new experiment to try to get the same result from a different source of electricity. In a somewhat grotesque echo of Franklin's lightning investigation, he set up the experiment that has come to define him in the public imagination. He hung flayed frogs by hooks from the metal railings of his terrace, their muscles connected to a long metallic wire pointing skyward as black clouds gathered and thunder pealed. Sure enough, the distant lightning had the same effect on the frogs dangling from the metal railing as did the artificial spark: their legs kicked a zombie can-can. (Decades later, this led to Galvani's enduring nickname of "frog dance master.")

He decided that due diligence required him to do the same experiment on a clear day. Despite it being a sunny day, every now and then, the frogs' legs would kick anyway. Galvani checked the sky. No sign of "stormy atmospheric electricity." Galvani approached the frogs. After watching their contortions for some time, he began to realize that their jiggles coincided not with storms, but rather with the movements of the brass hooks clanking against the metal railing. He walked over to a frog and pressed the hook from which it was suspended against the railing. The frog's leg contracted. He let go. The frog's leg went slack. He pressed again, and again, and each time he did so, the frog legs responded as if on command.

The fact that it jumped whenever the hook was manipulated suggested that there was something inside the frog itself, maybe a kind of lightning all its own. Or a Leyden jar, as Galvani speculated later. This could change everything.

Galvani spirited the frogs down to his lab, now seeking escape from any trace of distant lightning, which was what he assumed was inflaming their nerves just as the distant spark in his prior experiment had done. He laid one, still impaled on its hook, onto a metal plate, far from any electrical machine. The leg jumped. You can feel the tension and excitement in Galvani's manifesto as he describes this experiment. There was no possible source of external electricity—he had removed all of them. This could mean only one thing: proof that the electrical impulse was coming from inside the animal itself. Or, as he put it, the mechanism that allows the body to act "at the direction of the soul." It's the first time in the document—after pages cataloging his many experiments—that he dares to spell out the phrase "animal electricity."[20]

But he didn't publish right away. Scientist, Catholic monk, and Galvani biographer Brother Potamian ascribed this to his solid character: "He had not that intense desire for publicity that causes smaller men to rush into print with their embryonic discoveries the moment they get their first distant glimpse of a new truth."[21] It took another half decade before he had satisfied himself that there could be no other explanation for the phenomenon. In January 1792, Galvani published his results in a fifty-three-page letter he entitled "De viribus electricitatis in motu musculari" ("On the effect of electricity on muscular motion"). It appeared in *Commentarii*, the official publication of the Bologna Institute of Sciences, printed in Latin and disseminated only to a small circulation. Still, it spread like wildfire. Historians believe Alessandro Volta obtained an early copy[22]—it would explain why he was able to undermine it so quickly.

The electrician with the ambition

Alessandro Volta's circumstances were not entirely unlike Galvani's. He had grown up in Como, a small city in Lombardy on the shores of Lake Como, where his family were minor nobility.

The Volta money came from land and property revenues, and Alessandro and his brothers had inherited a fair bit more from a wealthy relative. The family owned several estates in Como and Milan.[23] Volta could have just enjoyed his money and indulged his curiosity as an amateur natural philosopher, as was fashionable at the time, but he chafed at the prospect of an obscure existence of provincial comfort. While he formally adhered to Catholicism, his first priority was ascension into the ranks of the natural philosophers, whom he venerated as the heralds of a new enlightened age. "The new age is exploding 'blind superstition' and the people's deliria of old times," he wrote when he was sixteen, in a bombastic paean to science.[24] Echoing the general contempt for theoretical physiology—with its animal spirits and nervous juices—Volta hailed the physical sciences, with their testable hypotheses, as "the useful sciences."

In particular, the emerging science of electricity appeared to him a manifestation of the triumph of the Age of Reason over superstition. In his view, Franklin's proof that lightning was an electrical phenomenon, for example—not caused by "the element of fire" as the ancient superstitions had it—showed that modern natural philosophers had unquestionably established their superior understanding of the world. Volta wanted more than anything to rise to the rank of natural philosopher, but not just an academic "letterati." Volta coveted the title of electrician.

He devoured the readings of the stars among them: Franklin, Musschenbroek, and Giambattista Beccaria, who along with Bassi had introduced Franklin's ideas to Europe. To infiltrate this eminent clique, Volta took an unusual approach: he began to write to them. Often. At the time, addressing leading figures like these without credentials or connections was considered quite bold. He was only eighteen, and yet was inviting comments on a juvenile theory of electricity, as if he were a professor casually engaging in collegial chat with equals. Eventually, he posted his lengthy thesis to Beccaria.

It took Beccaria a year to respond, and when he finally did, the sole missive was a print of a recent paper, in which he had laid out *his* latest new theory of electricity, a tortured derivation based on the friction of different substances and their respective inclination to "give" or "receive" electric fluid. However, his hypothesis had been politely ignored by other influential electricians of the time. This had probably smarted Beccaria enough, but to then be faced with the gall of a young upstart Volta pointing out that it disagreed with his own (totally uncredentialed) new theory of electricity would have been the final straw. After a few more fruitless exchanges, a clearly affronted Beccaria "invited" Volta to "maintain an everlasting silence on the topic of electricity."[25]

Volta gamely turned the subject of subsequent letters to other topics, but inwardly seethed at the disdain. So when he brought up his theory with another member of his rapidly growing network of aspirational correspondents, he was open to any suggestions. Paolo Frisi—who shared Volta's generally underwhelmed reaction to Beccaria's idea—advised Volta that, rather than trying to engage with him in yet more letters, he should "place as much emphasis on scientific instruments as on controversial theory."[26]

By that time, Volta was nurturing a new ambition: to become not just an electrician but a professor of electrics. For that, he would first need to become famous. He got to work on a new apparatus that would cement his reputation by proving his theory about the role of attraction in electricity: the electrophorus, a new tool that provided a "perpetual" source of electricity. Perpetual was maybe too strong a word, but in a considerable improvement on Leyden jars, the electrophorus could deliver 100 zaps before it would need to be recharged, and you could even use a Leyden jar to recharge it instead of faffing around with amber and silk.

"Superb and useful" were the words lavished on Volta by Carlo Firmian, his most important political patron at Pavia. He did not hold back: "Doing honor to your Country, and to all of Italy, Mother of the Sciences and the Arts." A few months later, at the age of thirty-four, Volta was chair of experimental physics at the university, and yet he still hadn't attained the level of respect he so longed for.

One reason for this was that two other experimental philosophers had invented something rather like the electrophorus a few years before, and it strained credulity to imagine Volta had never heard of it or them. Such suspicions were not assuaged by the fact that Volta—always more of an instrument man than a theory man—could never satisfactorily explain *how* it worked, or by what laws it was governed. When confronted, he hummed and hawed about producing a paper—but while he (very slowly) worked on writing it up, he realized that, actually, he might never need to publish the thing. What actually mattered was that the invention had already buoyed his reputation as an electrician. Thanks to the well-connected social and professional network Volta had fostered, the electrophorus was sent to electricians in cities from London to Berlin to Vienna. Apart from a few unappeasable dart-throwers, most other electricians didn't care about a theory as long as it yielded a tool that helped them do better science. But while some of those were now calling him "the Newton of electricity," the dart-throwers never entirely went away, sneering at his flimsy paper—which he had published despite its continued lack of convincing explanation[27]—and continuing to keep alive the low-key rumor that he had stolen the credit for the invention of the device. He couldn't shake it off for sixteen years, not even through his invention of a truly game-changing tool, the condensatore. This could *detect*, not generate, electricity. It was the most sensitive detector ever built.

Yet still his critics derided him as the inventor of "amusements electriques."[28] This was the moment that, in 1791, Volta—defensive, prickly, and a bit piqued—first read a copy of the *Commentarii*.

Volte-face

Initially, Galvani's manuscript delighted Volta. Though the electrician should have been put off by his prejudices against physiologists, when he repeated Galvani's experiments for himself, he became convinced. That spring, he enthused that "I have changed my mind [about the idea of animal electricity] from incredulity to fanaticism." Immediately, he wrote a paper in response to Galvani's manuscript, introducing it in the spring of 1792 as "one of the great and brilliant discoveries, which deserves to be considered defining an era in the physical and medical sciences." Concluding his paper, Volta wrote that Galvani has "all the merit and paternity of this great and stupendous discovery."[29]

But this full-throated approval would not last. By his next publication, a mere fourteen days after the first one, Volta's ardor had cooled substantially.[30] He casually put forward an alternative explanation for the frog-leg contractions—it was the metals Galvani had used, he said, that were solely responsible for the electric charge—and accused Galvani of being ignorant of some fundamental laws of electricity. Volta had seen how materials could respond to a far-off electrical source without contact being necessary. Perhaps, he began to wonder, if Galvani had been aware of this law, he would have correctly identified the material on the hooks as the cause of the contraction rather than some electricity intrinsic to the frog.

ARTIFICIAL vs. ANIMAL

Volta wasn't the only one who went from hot to cold. The Italian physician Eusebio Valli visited the French Académie des sciences to demonstrate Galvani's experiments there.[31] Valli had been among the first to publish a supporting paper on animal electricity, in which he wrote that "Galvani's discovery" had robbed him of "sleep during several nights." After witnessing the demonstrations, the Académie launched a series of replications, their usual approach for putting promising or controversial research through its paces.[32] It appointed several established scientific authorities to the commission, among them Charles Coulomb, a French physicist who would go on to describe the electrostatic force of attraction and repulsion, and whose name is now synonymous with the standard international unit of electric charge. However, the commission's eagerly awaited findings never materialized. The science historian Christine Blondel points to "uncertainty about the theoretical interpretation" Galvani lent to his experiments: code for a suspicion on the committee that Galvani was simply dressing up old superstitions as nouveau science.[33] In any case, the report disappeared and the Académie remained noncommittal.

Volta had no such reservations—he had done more replications of his own, and had begun to suspect Galvani was badly misconstruing his own results. The problem was this: when Volta did the experiments, the frog's muscles didn't always contract. Sometimes they did, sometimes they didn't, and Volta thought he saw a pattern emerging. When he connected the frog elements using a wire made of two different metals (for example, tin and silver), the legs could be trusted to jump. But if he used a wire made of only a single metal? The frog legs were as likely to twitch as to remain lifeless. This pattern led Volta to suspect that maybe Galvani was looking at the experiment backward—instead of springing from some inherent biological electrical flow within the frog, maybe the electricity had been entering the frog from outside all along. Maybe it was something about the metal in the wires that was actually generating the electricity.

Still stung by the fact his electrophorus had earned him a professorship but not philosophical acclaim, Volta continued his hunt for a general theory of electricity to cement his reputation as a brilliant theorist—and thought he had found it in Galvani's misconstrued results. Six months after the publication of Galvani's "De viribus," Volta published this alternative explanation of the contractions. First he aggressively debunked Galvani: "equating the animal spirits with the electrical fluid that flows through the nerves is one of those 'plausible and seductive' explanations that have to be withdrawn in the face of contrary experiments," he wrote.[34] In his view, the contracting legs actually demonstrated the power of the "dissimilarity of the metals" in the wire that had been inserted into the frog. After all, if the reason for the frog's jerking leg had simply been unbalanced animal electricity, the composition of the wire that connected the frog's limbs should have made no difference to the results. But it did matter, as Volta's own experiments showed. To get a sure twitch, you needed a wire made of "two metals that are of different kinds or that are dissimilar in some other way, such as in hardness, smoothness, shine, etc," he wrote.

Volta hypothesized that contact between any two different metals automatically generates electricity all by itself. Metals, he said, "should be regarded no longer as simple conductors, but as true motors of electricity, for with their mere contact they carry it around."[35] As his confidence in this explanation grew, his language grew more aggressive. "There is surely no reason to assume that a natural, organic electricity is at work here," he wrote in one paper. In an open letter published at the end of the same year, he dropped the gauntlet. "If that is how things are, then what is left of the animal electricity claimed by Galvani? The entire edifice is in danger of collapsing."

Many undecided scientists were swayed by these muscular proclamations; Galvani's frogs were in hot water. In response, Galvani produced a new experiment. Volta retaliated with one of his own. And so it went: experiments and counter-experiments, each designed to prove conclusively that the other was wrong. Nevertheless, the two (largely) persisted in gentlemanly conduct: as late as 1797, when the differences in their interpretation of the frog experiment had become insurmountable, Galvani still emphasised Volta's "erudition" and "depth of wit," and Volta in turn called Galvani's experiments "very fine."

The same could not be said of their contemporaries, who had long since split into bilious factions engaged in proxy combat. Volta's declarations were made "with the thunder of truth," according to the physician Giovacchino Carradori. The chemist Valentino Brugnatelli bombastically announced "the ruinous downfall of Galvani's theory" under the "repeated attacks of a terrible adversary." One of Galvani's most loyal supporters was his nephew Giovanni Aldini, who had not only helped with the experiments, but had written some of the publications himself. He was incensed by what he perceived as the baseless attacks. "If the good repute and integrity of scientific opinion were to be called into question whenever the slightest doubt was raised, we would certainly have few or no theories," he sniped in a letter to Volta.

As for Galvani himself, he unwaveringly disputed Volta's characterization of his inability to provoke contractions from a single metal: "I can assure you that I obtained the motions not a few times, as Volta claims, but in many, many experiments, so that in a hundred times the effect had not happened just once," he explained to his old friend Lazzaro Spallanzani. "These experiments have been recently replicated by other people well versed in this sort of thing, and they never failed." The variability, he explained, was largely a consequence of other researchers using frogs that had been dead for longer than forty-four hours. In addition, they had not necessarily followed Galvani's exacting means of preparation.

By now, so many scientists had joined the cause that Europe started running out of frogs. "Sir, I want frogs," Valli admonished a colleague when he ran out while replicating one of the experiments. "You must find them. I will never pardon you, sir, if you fail to do so."[36]

All the while, no one could reach a definitive conclusion on the validity of the animal electricity that was increasingly being referred to as "galvanism." After the first French commission of the Académie des sciences ended in uncertainty, in 1793 the baton passed to the Société philomatique de Paris, an entity founded with the explicit mission of "repeating doubtful or little-known experiments." Instead of great physicists, however, the Société had the second commission run by three amateur scientists.[37] Though they seemed less hostile toward Galvani, again they were unable to offer a definitive verdict on galvanism.

By then, in 1794, Galvani was ready to claim victory for good. He understood that if he was to prevail, he would need to prove that it was possible to obtain contractions without the aid of any metals at all; if he could get the same dancing-legs result sans wires, Volta would have to concede. So he did just that: after a grueling series of variations on the original experiment, he was finally able to subtract the confounding wire, instead managing—with an anatomist's delicate precision—to surgically connect a frog's muscle directly to its nerve. The leg jumped.

Here it was at last: irrefutable proof that intrinsic electricity coursed through animal tissues—its vestiges remaining at least for a while after death—entirely isolated from any possible external source of metallic electricity. He had long thought that a muscle was like a Leyden jar, whose spark could be freed by a conductor—and here was proof that in animal tissue, the nerves were the conductors. Galvani published. His powerful and loyal friend Lazzaro Spallanzani lent the weight of his reputation, proclaiming that he had succeeded in "confuting victoriously the objections."

Now everybody wanted to be a Galvanian. Valli declared victory on behalf of Galvani, saying that "metals possess no secret magic virtue." Membership expanded: "Thunder of truth" Carradori abandoned Volta for his rival, as did "ruinous downfall" Brugnatelli. (Indeed, in the wake of the third experiment, Brugnatelli now claimed he too had obtained movement in his frog "without the help of the metals."[38]) Galvani's relief was almost palpable in a letter to Spallanzani shortly afterward, thanking him for the support. "It could not be more courteous and appreciated," he wrote. "This letter produces a fulsome calmness in my soul, which was indeed rather restless."

Galvani and his supporters were convinced that the new results would finally put an end to the controversy. A rumor even spread that in December 1794 Valli had met Volta in Pavia and "converted" him. The rumor was unfounded and Volta was furious. Immediately he set about writing a series of letters to Anton Maria Vassalli, the secretary of the Turin Academy of Sciences, dissecting Galvani's latest publication and the social fallout it had occasioned. "These experiments impressed many people, and drew them toward Galvani's banners when they had already subscribed, or were going to subscribe, to my totally different conclusion." Volta could not be right if Galvani was not wrong.

In his letters with Vassalli, Volta set out his riposte. Perhaps, he theorized, the connection between the muscle and the nerve wasn't a home run for "animal electricity" after all. Because what if—like metals—different kinds of tissue also allowed a very thin charge to pass between them as long as they were sufficiently dissimilar? In other words, maybe nerve and muscle were just the biological version of tin and silver; their differences, put into contact, would cause electricity to flow.

This insight inspired him to return to the discovery that had led him to examine the difference in metals in Galvani's initial experiments in the first place: the theory of dissimilar conductors. He decided to broaden his theory of metallic contact beyond metals. "Any time two different conductors are connected, an action arises, which pushes the electric fluid," he announced. As long as a circuit is closed, and as long as the materials are very different, "some current is constantly excited." Even meat could be a material that conducts, as long as it is joined to another kind of meat that is sufficiently dissimilar. Once again public opinion swung in Volta's favor.

After months of trying to figure out how to connect two such gossamer fibers, Galvani suddenly realized what he had to do next: connect two nerves within the same frog, instead of connecting a muscle to a nerve. He aligned the cut end of a frog's left sciatic nerve with its right sciatic nerve and then aligned the cut end of the right sciatic nerve with its left sciatic nerve. It was exactly the same kind of tissue within the same animal. No conceivable difference, metallic or biological. And still: both legs jumped.[39]

With that, he undercut Volta's last remaining objection to the idea of an innate electrical current inside an animal: according to Volta's own logic, two nerves, comprised of exactly the same material, should not possibly generate any charge at all. Which meant there could be no other explanation of the current being witnessed in the nerves—it had to have a physiological origin. Galvani sent his manuscript to Spallanzani in 1797, whose response was unreserved. "For [its] novelty, for the importance of its doctrines . . . for the clarity and brilliance with which it is written, this work appears to me as one of the most beautiful and valuable of the eighteenth-century Physics," he declared. "With it, you have erected a building that, because of the firmness of its foundations, will last for the centuries to come." It was a prescient statement. The series was a fundamental experiment for the foundation of

ARTIFICIAL vs. ANIMAL

all electrophysiology. Neither Volta nor the other adversaries of animal electricity ever bested it.

This should have ended all argument. Galvani should have harvested the fruits of all his long years of experiments. In a just world, Galvani would have been showered with awards and honors, and his success would have led to an enormous boom in electrophysiological research focused on narrowing down exactly what kind of electricity it was that flowed through the nerves.

But that's not what happened. Instead, Galvani's beautiful coup de grâce went practically unnoticed by the scientific community and was nearly lost for ever. That's because Volta was about to unveil his world-changing instrument: the battery. Volta had been busy turning his expanded general theory of contact electricity into a physical device. According to the theory, the frog in Galvani's original experiments had simply functioned as a damp material that closed the circuit between two dissimilar metals—a "moist conductor." So why not create an artificial "frog" but with wet brine instead of the wet frog?

Sure enough, Volta found stacking two discs of different metals, separated by a cardboard disc soaked in brine and connected at both ends by a wire, produced a spark. The higher you piled the discs, the bigger the spark. This cemented Volta's conviction that Galvani had his hypothesis backward and helped him sell his version of the story to other scientists. All Galvani had really done, Volta claimed, was to create a semi-biological version of his "voltaic pile," in which the brine was replaced by the rather more cumbersome frog. Remove this overcomplication, and you got a device that was able to store and release continuous charge—in other words, a rudimentary battery.

The final blow to Galvani's place in history was delivered not by science, but by politics. Bologna had succumbed to the French occupation of northern Italy. Napoleon's Cisalpine Republic insisted that every university professor must swear an oath of loyalty to its authority. By 1798, Volta and Spallanzani had taken the oath, but Galvani was still a holdout.[40] He could not bring himself to make such a concession to an authority that conflicted so heavily with his social, political, and religious ideals. "He did not believe he ought to, on so serious an occasion, permit himself anything but the clear and precise expression of his sentiments," wrote his first biographer, Giuseppe Venturoli, a fellow professor at the University of Bologna during the galvanism wars, who had remained an unwavering Galvanian. He also refused to take advantage of the suggestion that he should modify the oath by some subterfuge that betrayed his principles. The price for his refusal was steep; he was stripped of all his academic positions, leaving him without income, estate, or purpose. After lengthy consideration, in 1798, the republican government decided to overlook the refusal and reinstate him. But the decision came too late: by the time the word was handed down, he was already dead.

The urgency of finding what he conceived of as God's "breath of life" had kept Galvani toiling through countless hours in a lab surrounded by the corpses of dead frogs, through the heartbreak of his wife's death, and through an excruciating public attack on the validity of his scientific discoveries. But a man has his limits. Luigi Galvani died in his brother's house in Bologna, poor, anguished, and stripped of his titles, on 4 December 1798.

By the time Volta formalized his victory in 1800 by publicly showcasing his demonstration of the voltaic pile to the president of the Royal Society of London, word of the stupendous new invention had spread widely—he had been writing drafts since 1797, and had certainly shared them with colleagues. He had thoroughly won the day. The battery invalidated Galvani's claim to the existence of animal electricity—not because Volta had proved it, but because he said so.

Apart from a few stubborn Galvani loyalists like Spallanzani, the voltaic pile swung the scientific community to his side. "Thunder of truth" Carradori switched teams one last time to back Volta, along with "ruinous downfall" Brugnatelli.[41]

With no leader left, the serious study of animal electricity fell away. Neither Galvani nor his supporters had ever been able to measure animal electricity with any kind of electrometer. The current was simply too faint to be detected by the instruments of the day. No instrument had emerged from the ruck of studies—French and otherwise—that could support the theory of animal electricity the way the obviously useful voltaic pile had immediately buttressed Volta's notion of metallic contact electricity. Volta could prove his theories with a tool and many use cases. Galvani could not.

One crucial limitation of Galvani's experiments was that he was never able to separate the source of animal electricity from its detector—they were both the frog. No similar confusion plagued Volta's research. That put Galvani at a major disadvantage, because it muddled the terms.

So while Volta's invention of the battery did not itself invalidate any of Galvani's theories about animal electricity, it effectively shut down all further challenge. Volta had changed the terms of the debate, leaving his contemporaries so dazzled by the device and its potential that they forgot what the original fight was about. Galvani's ideas weren't so much disproved as abandoned.

The long tail

In the wake of Volta's perceived victory, Galvani's theories were shunned by science for nearly half a century. Galvanism was quickly overrun by quacks and their most gruesome pseudo-medical treatments. At the same time, the battery—and the "artificial" electricity it was able to make flow continuously for the first time—quickly went on to underpin many of the century's most important advances in the physical sciences. It allowed Michael Faraday to come up with the laws of electromagnetism, and in more practical terms, it powered telegraphs, electric lights, doorbells, and eventually power lines. In the hands of the physicists, artificial electricity transformed civilization.

If the Galvani–Volta battle set the stage for the separation of what we understand today as biology and physics, it was only the beginning. Eventually, better tools were able to detect the exquisitely faint electrical currents running through frogs' legs, but by then it was too late. The idea had been set: electricity was not for biology. It was for machines, and telegraphs, and chemical reactions. Not until the following century did research on biological electricity return to being a legitimate scientific pursuit, and even then, it returned in a much more circumscribed context.

The historians Marco Bresadola and Marco Piccolino note that outside Bologna, even two centuries after his death, Galvani's contribution to science was still represented mainly as that of a know-nothing anatomist whose accidental insights helped midwife Volta's invention of the battery. But the person who cemented that reputation immediately after Galvani's death wasn't Volta—in fact, it was the last person you might have expected.

CHAPTER 2

Spectacular pseudoscience: The fall and rise of bioelectricity

Giovanni Aldini was looking for the perfect body. Not something that had been dragged out of a grave—it should be as fresh as possible to minimize the dissipation of its vital powers. It shouldn't be someone who had died slowly, from one of the "putrefying diseases" that might contaminate their humors.[1] Not too dismembered, either. The ideal body would be one whose previous owner had been healthy and intact until the moment of death.

Aldini's star had been rising in Europe as he demonstrated Galvani's experiments on much larger animals than frogs, to often macabre effect. In an echo of some of the early electricity spectacles—but with a darker twist—he had recently electrified a decapitated dog to entertain a crowd that included royalty.[2] He was desperate to prove that the animal electricity Galvani had discovered was present in the same way in all animals—that what was true for frogs was true for humans. He was willing to use Volta's battery and any amount of theatrics to prove it.

Aldini was in the right place at the right time: it was 1803 in the UK, and the Murder Act had for well over half a century included a provision that would serve up exactly the corpse he was after. After a convicted killer's public hanging, their naked body would be flayed in a public dissection. If that seems over the top, it was absolutely intended to be—this "further terror and peculiar mark of infamy" had been added to give would-be murderers just that extra bit of pause, all the better to prevent "the horrid crime."[3] It was unclear whether, as Aldini would later write, it also better helped them atone for their sins, or whether there was a more convenient secondary benefit; as there were laws against digging up corpses, this law provided a steady stream of cadavers to upskill medical students and lecturers at the Royal College of Surgeons.[4] The fellows there had invited Aldini from Italy to demonstrate the experiments that had recently made him famous around Europe.[5] They were happy to supply the necessary materials. And so, after the convicted murderer George Forster was hanged on the gallows at Newgate prison, his body was carried across town to the Royal College of Surgeons, where Aldini nervously waited.

The room was crowded with luminaries, scientists, and gentlemen standing elbow to elbow. Assisting Aldini in his efforts would be the rising star Joseph Carpue, a surgeon and anatomist at the Duke of York's Hospital who had extended the invitation, and Mr. Pass, the beadle of the Surgeons' Company, who was tasked with making sure all proper protocols were followed during a dissection.[6] But the crowds weren't what had Aldini sweating; he was used to performing in front of high society.

What was worrying him today was the cold: it was January, and the body had been left hanging for an hour in temperatures two degrees below freezing. The chill might stunt the flow of animal electricity through the body, rendering his experiment a humiliating, public flop. He was putting his faith in the enormous piles of alternating zinc and copper discs sitting on the slab where Forster's corpse was now laid out, ready to dispense their "galvanic juices" into the dead man's nervous system.

Aldini moistened the tips of two metal wires attached to either end of the pile by dipping them into saltwater. When he threaded them gingerly into Forster's ears, the results did not disappoint. The dead man's jaw, according to a report in *The Times*, began to quiver: "the adjoining muscles were horribly contorted, and the left eye actually opened," giving the impression of a ghastly, lewd wink.[7] Over the next several hours, Aldini's team exposed every nerve and muscle on the man's body, from the thorax to the gluteus, for electrical experimentation.

Forster wasn't Aldini's first criminal corpse. He had spent the previous year in Bologna and Paris perfecting his galvanic technique on the heads and bodies of other hanged and beheaded convicts, not to mention the scores of lambs, dogs, oxen, and horses, living and dead, that joined Italy's population of frogs on his table. These animal experiments had given him the idea for an especially dramatic demonstration.

When Aldini plugged one of the wires into the dead man's rectum, the convulsions that wracked the corpse were "much stronger than in the preceding experiments," Aldini wrote. So strong, in fact, "as almost to give an appearance of reanimation." At this point, according to *The Times*, "some of the uninformed bystanders actually thought the wretched man was on the eve of being restored to life." Some in the audience clapped; others were deeply disturbed. Mr. Pass was so shaken by what he saw on the table that he went home that night and died.[8] As far as Aldini was concerned, the experiment had been a success.

This spectacular public demonstration begat many copycats, and historians trace a line from the Forster galvanization to Mary Shelley's idea for *Frankenstein*. So it may come as a surprise that Aldini didn't start out with the goal of titillating vapid royalty by raising the dead. He was pushed down his path by an altogether more noble impulse: to restore the reputation of his beloved uncle. But not unlike Dr. Frankenstein, his obsession caused him to reach beyond what science could provide, and eventually turned him into a mockery. He would become a scientific pariah. Instead of reviving his family's legacy as well as decapitated bodies, his experiments would play a major role in banishing the serious study of animal electricity into a desert of quacks and mountebanks for the next four decades.

Aldini's gambit

Aldini's loyalty to Galvani wasn't just a matter of family honor. He had also been his uncle's closest and most important scientific collaborator. He had written some of the anatomist's famous communications himself—some of the liveliest ripostes between "Galvani" and Volta had actually involved just Volta and Aldini.[9] But after Galvani's death, few champions remained to take forward the serious scientific inquiry into animal electricity.

In 1801, Napoleon's French Académie launched a commission (the fifth in as many years), offering a prize of 60,000 francs for anyone who could do for animal electricity what Volta had done for the metallic or artificial variety.[10] (In today's money, this would have been worth around £860,000.) Generous as it was, however, the prize went unclaimed. No one was in the position to make something as consequential as a battery for animal electricity. In addition, the false perception that acceptance of metallic contact theory and animal electricity had to be mutually exclusive meant, for many, that because Volta (so heavily favored by Napoleon) had been demonstrably right, Galvani must by definition have been wrong.

Aldini was desperate to stop this becoming the official received wisdom. He had understood the scientific foundation his uncle was trying to build, and he had noted the sleight of hand that undermined it. In particular, Aldini was still pained that their most triumphant paper—the one Spallanzani had hailed as "one of the most beautiful and valuable of the eighteenth-century Physics," in which Galvani had showed up Volta once and for all by successfully proving that nerve electricity could excite nerve tissue—was already being forgotten. This was the paper that should have put the lie to Volta's insistence that the only reason contractions could be evoked in a dead frog was that some version of metallic electricity was generated by the meeting of two dissimilar kinds of meat. Instead, the paper had been buried under the fanfare around the voltaic pile.

And so, Aldini's initial investigations after his uncle's death focused on buttressing the basic science underlying this experiment, and how it could advance a deeper understanding of animal electricity. He had assumed the chair of physics at Bologna in 1798, just before Galvani passed away. This was a prestigious post from which to carry on his uncle's work, and Aldini used it to launch the Galvanic Society of Bologna.

Galvani had experimented almost exclusively on frogs. Aldini's first experiments therefore extended his uncle's investigations into warm-blooded mammals. His 1804 publication *Essai théorique et expérimental sur le galvanisme* is filled with long, repetitive accounts of experiments in which he and his Galvanic Society collaborators sought to understand "intra-animal" electrification. In one characteristic experiment, he placed several calves' heads in an electrically conducting line called a "series," and used the resulting animal current to violently electrify a dead frog. But when he tried to reverse the experiment, applying the animal electricity of frog nerves to the decapitated heads of oxen, he found the results less dramatic and even disappointing. All these experiments successfully replicated Galvani's original idea that the same electrical substance coursed through all animals, but none yielded grand dramatic outcomes or novel insights.

At some point, it seems to have become clear to Aldini that to maintain excitement about scientific galvanism, he would need to do what the five commissions hadn't been able to: find a way to make his uncle's discoveries medically relevant for humans. It was around then that his focus rather quickly shifted, revealing a sudden new appreciation of the "galvanic juices" dispensed by Volta's pile. "The battery imagined by Professor Volta gave me the idea of a cleaner means than any of the ones we have used so far to estimate the action of the vital forces," he recalled in the 1804 *Essai*.[11]

It must have been hard for Aldini to hold his nose and use the instrument of his uncle's doom, but once he got the hang of it, he was prolific. He used the pile's ability to dispense a steady flow of electricity to stage big, dramatic experiments on dead animals. He inserted wires into their rectal cavities, often detailing the inevitable violent expulsion of feces that followed. He also began to experiment with touching different areas of the animals' brains, as well as his own; when he administered a jolt from the pile to his own cranium, it led to a few days of insomnia but also a strange feeling of elation.

Such experiments fascinated the other members of the Galvanic Society: if a jolt to the head could make Aldini feel euphoric, what else could it do? They analyzed and repeated these kinds of experiments until they eventually accreted into new theories about how electrical ministrations could improve ailments. Most promising were epilepsy, a type of paralysis called chorea, and what was then called "melancholy madness," which we understand today as treatment-resistant depression. Now they just needed test subjects.

In 1801, at Sant'Orsola Hospital in Bologna, Aldini found a twenty-seven-year-old farmer called Luigi Lanzarini who had become catatonic with melancholy madness and had been declared a lost cause.[12] He shaved Lanzarini's head and stimulated the man's scalp with a weak battery. Over the next month he slowly increased the current, and Lanzarini's symptoms seemed to lessen, eventually enough for him to be released into Aldini's custody. After about a month, Aldini deemed him well enough to send back home to his family.

Word of this achievement spread quickly enough that, by 1802, French scientists founded their own Parisian branch of the Galvanic Society. They devoted themselves to Aldini's goal of elevating the reputation of galvanism as a legitimate pursuit, by any means necessary. Joseph Carpue—the rising-star surgeon who had assisted in the Forster experiment—reported that a M. La Grave of the Parisian Galvanic Society had made a voltaic pile out of sixty layers each of human brain, muscle, and hat material (you read that right) moistened with salt water.[13] Its effect was allegedly "decisive"—generating a current that provided yet another piece of evidence that animal electricity was just as relevant and present in human tissues as it was in animal tissues.

It was never entirely possible to extricate galvanism from its associations with woo and quackery—"a couple of [the Galvanic Society's] members drifted into 'galvanic magic,'" notes the historian Christine Blondel—but most of the group's research was greeted with interest by French and foreign scientific journals and even encouraged.[14] The attention-seeking experiments were doing their job. Famous French psychiatrists began to consult Aldini about use of the pile to restore their patients to health.

But by then, Aldini already had his eye on an entirely new patient population: he started investigating electrification as a way to revive the dead.[15] To be clear, his goal was never to stitch together some kind of undead golem—Aldini was referring to the evidently reversible state of "suspended animation" that followed accidental drowning, apoplexy, or asphyxia.[16]

Aldini was campaigning to get galvanism—specifically a jolt of electricity to the head—included in the go-to methods for emergency resuscitation, which included ammonia and a kind of proto-CPR that involved exhaling into the lungs of the temporarily deceased.

Adding an electric shock to either of these remedies, Aldini insisted, "will produce much greater effect than either of them separately." He also began to lobby to have electrification adopted as a research tool to determine whether someone was truly, irreversibly dead. "It would be desirable to establish by public authority, in all nations, by people enlightened and able to make the necessary tests to determine whether death is real or not."

Of course, it's well known today that his hunch was correct—electrical defibrillation can bring a person back from what would have been certain death. But Aldini's speculations were not based on any specific mechanism or evidence. He had no access to any of the information that we take for granted 200 years later: that a meaningful resuscitation is largely determined by whether or not the patient is brain-dead; that it's crucial to keep oxygen moving to the brain; that there is a small window of time after which any attempt at resuscitation becomes useless. Unfortunately, even the most fundamental mechanism eluded Aldini—that the organ that should be stimulated is the heart, not the brain. In fact, he repeatedly and explicitly refuted the idea that the heart could be affected by electrification at all. His focus on spectacle over basic science had misled him.[17]

So it came as no surprise to him that none of his experimental subjects—human or animal—were ever shocked back to life. Neither was such an outcome ever his goal for the hanged Forster. "Our object here was not to produce reanimation, merely to obtain a practical knowledge of how far Galvanism might be employed as an auxiliary to other means in attempts to revive," he wrote in an 1803 account of the experiment. This writing also provides clues to how he thought galvanization might act to restore the dead to life: by "re-establishing the muscular powers which have been suspended," in addition to preparing the lungs to receive resuscitation.

However, these were not the promises that got royalty crowding around his table. It was the extras: the grimaces, the rectal probes, and the unspoken possibility that, just maybe, one of the malefactors might rise from the dead. Word had begun to spread of his work on deceased criminals in Bologna in early 1802.[18] He had managed to raise the forearm of a corpse to the height of eight inches seventy-five minutes after death, "after we put in his hand a fairly considerable weight, such as an iron pincer." Stimulation of the arm caused the hand to rise and curl into a position that looked for all the world like an accusing finger pointing at the assembled audience, several of whom promptly fainted. His colleagues at the Galvanic Society, Professors Giulio, Vassalli, and Rossi, quickly repeated these experiments in Turin on three recently decapitated men.[19] It wasn't long before such demonstrations piqued the interest of the Royal Humane Society of London, though maybe not for the reasons you might expect.

These days, a person who identifies as humane might find cause for concern in the public dissection of dead criminals for entertainment. Not these officials. They had more pressing concerns, like how to distinguish between people who were genuinely dead and those who were going to wake up.[20] Before the wide availability and awareness of reliable methods of resuscitation, burials could be a pretty hasty affair, and more than one unfortunate had found themselves waking from a comatose or cataleptic state (or just a deep and drunken sleep) inside a little box under six feet of earth. Sometimes, their screaming was heard in time. (In one especially grisly case, this fate befell the same poor woman *twice*.) "A host of facts have repeatedly shown us that people were rushed to the tomb before death had irrevocably struck them. Shouldn't we give our fullest attention to preventing such deadly events?" Aldini had written, scandalized by the haunting stories of these potentially "murderous burials."[21] Busy, commercial, and maritime Britain was seeing a glut of drownings and mining accidents, and so for the Royal Humane Society, having some way to distinguish the dead from the "not actually dead even though they sure look like it" was very much at the top of the agenda.

In late 1802, they sponsored Aldini for a long tour of Oxford and London, and that's how he came to be in the room with Mr. Pass and Mr. Forster that chilly morning. Did he think the man would wake up there on the table? Certainly not. Did he think the experiment would contribute to better resuscitations? Sure—but there is little evidence in his writing of any empirical understanding of how the stimulation might accomplish that. So in some way he must have understood that what he was doing there that day was to a large extent more showmanship than science.

Unfortunately, in trying to preserve what was left of his uncle's nascent science, Aldini failed. He did, however, have great success at blurring the line between "legitimate" galvanism and the unscientific electroquackery that had begun to proliferate long before Galvani touched his first frog. And the quacks came marching over it.

Elisha and the quacks

Almost as soon as the Leyden jar was invented in the mid-1740s, people were convinced its zaps could dispense powerful cures.[22] In Italy, its invention prompted the opening of no fewer than three schools of electrical medicine. Treatments there varied—some doctors simply shocked their patients and hoped for the best; others hoped that electrical stimulation would boost topical medicines' ability to reach deep beneath the skin. The practice was said to cure a spectrum of ailments so wide it bordered on the miraculous.

No ailment was spared the intervention of the Leyden jar, including but not limited to gout, rheumatism, hysteria, headache, toothache, deafness, blindness, irregular menstruation, diarrhea, and, as ever, venereal disease.[23] By the 1780s, electricity was rumored to beget miracles, as in the report of a couple who, after ten years of infertility, "regained hope through electricity thanks to a few turns of the crank and some shocks in the appropriate parts"—the Abbé Bertholon, reporting this, "demurely did not specify which."[24] It wasn't just a continental fad: Britain's medico-electric quack culture also thrived, adding "weak ligaments," testicular and urinary conditions, and ague (that's "the shivers" to you) to the list of conditions electricity would relieve. It was hard to beat the electrical apparatus devised in 1781 by the London medical electrician James Graham, who guaranteed that his electrically stimulating Celestial Bed, in a special wing of his Temple of Hymen, would cure sterility and impotence.[25] What put this electro-quackery a cut above the usual sort was that no actual electricity was involved—merely the idea of it, as Graham figured "the fashion of electrical vapors" would suffice to cure his patients.[26] A night in this contraption would set you back £50, around £9,000 in today's money,[27] but if you still had money burning a hole in your pocket, you could hit the temple gift shop on your way out for a take-home aphrodisiac called Electrical Ether. (Given that the temple closed its doors within two years, "homeopathic electricity" does not seem to have been an unqualified success.[28])

But it was Galvani's science that would inspire the most brazen of these quacks: Elisha Perkins. "Among the delusions which have succeeded in imposing on men of education and position, it is pre-eminent," wrote Francis Shepherd, describing "Perkinism" in the *Popular Science Monthly* in 1883.[29]

Perkins—who at the time of Galvani's "De viribus" publication was practicing medicine in Connecticut—had keenly followed accounts of the fight raging on the Continent, and saw an opportunity in the argument about dual metals.[30] One that could make him rich. In 1796, he unveiled his contribution to medical galvanism—a pair of sharp-tipped three-inch rods, one iron, one brass, which he called "tractors." Drag them over your afflicted parts for a few minutes, and he claimed that you would soon be rid of rheumatism, pain, inflammation, and even tumors. Perkins's patented tractors took off in America among the wealthy and influential. Even George Washington bought a set for his family, along with Chief Justices Oliver Ellsworth and John Marshall.[31]

The Connecticut Medical Society was having precisely none of it. In a scorching rebuke of Perkins, they began expulsion procedures with a letter stuffed to the margins with indignation. Castigating Perkins's inventions as "delusive quackery," they accused him of using the auspices of his Society membership to spread his "mischief" to the south and abroad. "We consider all such practices as barefaced imposition, disgraceful to the faculty, and delusive to the ignorant," the Society fulminated. With that, they invited Perkins to "answer for his conduct, and render reasons why he should not be expelled from the Society, for such disgraceful practices."[32]

Whatever reasons Perkins may have offered, they failed to sway the opinion of the Society, which in 1797 expelled him for violating their prohibitions against nostrums (medical treatments prepared by unqualified chancers). This goes some way to explaining why Perkins's son soon took the family business to the Continent. They were a wild success. In 1798, the Royal Hospital in Copenhagen officially adopted the tractors for treatment. In London, the Royal Society "accepted" the tractors and the accompanying book (there's always a book), and by 1804, a Perkinean Institution had been established. Members included fellows of the Royal Society. Soon a hospital was built whose sole treatment was "tractoration." Testimonials abounded: not least from bishops and clergymen, whom Perkins had slyly provided with that most ancient grift: free review samples. "I have used the tractors with success in several cases in my own family," wrote one recipient, channeling the logic of a multilevel marketing scheme. "Since experience has proved them, so no reasoning can change the opinion."

Over time, galvanism was conscripted into a pre-existing, ever-widening gyre of pseudoscience that included Franz Mesmer's animal magnetism, hypnosis, and electrifying wearables, variously said to be associated with earthquakes, dowsing, and volcanic activity. The whole line of research was evidently starting to exasperate the public. In an 1809 poem, Lord Byron lumped galvanism in with tractors eleven years after Galvani's death, apparently channeling the public mood that had started to conflate both:

> What varied wonders tempt us as they pass
> The cow pox, Tractors, Galvanism and Gas
> In turns appear to make the vulgar stare
> Till the swollen bubble bursts—and all is air![33]

"A prostitution of galvanism"

In the end, Aldini's efforts to clear his uncle's name had the opposite effect. They created a self-perpetuating spiral that destroyed what was left of Galvani's reputation as the father of animal electricity: the more quacks who co-opted galvanism for their own purposes, the fewer legitimate researchers were willing to be associated with the relationship between electricity and life; the less serious research took place, the more ground was ceded to ludicrous claims. As the years went on, an increasing number of scientists and historians looking back on the Volta–Galvani feud began to ad lib historical details about Galvani that confirmed the cynical new perspective on animal electricity—and his ignorance in believing it to exist. One of the most enduring of these is the pernicious origin myth that Galvani accidentally bumbled upon the idea of animal electricity as his wife was preparing a frog for soup using a metal knife, rather than in a decade-long series of ever more finely honed replications.

At the same time, the sciences were now rapidly beginning to diverge into their fields, and biology was defining itself as a discipline. Not wanting to make Galvani's mistakes, those who pursued the legitimate study of biology recoiled from electricity and returned to a largely descriptive anatomical and taxonomic focus, a study of pieces instead of the forces and processes that govern the whole.

People who studied electricity seriously—the electricians—were desperate to restore respectability to their endeavors, and that meant separating the object of their study from vitalistic connotations, and focusing strictly on the advances being made by chemists and physicists thanks to Volta's battery. These multiplied quickly. In 1800, his proto-battery helped scientists electrolyze water, decomposing it into hydrogen and oxygen. In 1808, an improved version helped chemists to discover sodium and potassium and alkaline earth metals. Equations were devised to define the relationships by which electricity could act on the world. In 1816, the first working prototype of a telegraph was built in Hammersmith, powered by voltaic piles. The physicists and engineers had created an electric force field around themselves that no one could touch, protecting themselves from both the biologists and the charlatans.

Medical professionals also separated themselves from animal electricity in due course, even as some of them continued to deploy the artificial electricity that could zap people's ailments. In the 1830s, a young doctor called Golding Bird—after seeing the quacks make money hand over fist—set up "electric baths" facilities at

Guy's Hospital in London, where he charged his posh patients a hefty fee to alleviate vague maladies.

But not everyone abandoned the project of building a legitimate discipline around investigating animal electricity. Behind the scenes, another scientist had been working to keep its study on life support: Alexander von Humboldt had reviewed Galvani's work for the French commission throughout the 1790s, and had grown to strongly suspect that Volta and Galvani's theories did not contradict each other after all, and that Volta had in fact been wrong to dismiss animal electricity out of hand.[34]

Humboldt would go on to become chamberlain to the Prussian king and a leader of the Enlightenment, shaping how we conceive of nature itself as a single interconnected system. But during the electricity wars he was still in his early twenties, having recently graduated from university into a position as a mining inspector. Quickly his polymathic tendencies jumped from geology to botany to comparative anatomy. When he became aware of the Volta–Galvani controversy, he set his mind to solving the mystery.

To that end, Humboldt conducted about 4,000 experiments over five years, several on himself. (His friend Johann Wilhelm Ritter, whom he often convinced to join him, scrambled his nervous system so much with this type of self-experimentation that he died at thirty-four.) Of these investigations, arguably the most shocking was Humboldt's decision to insert a silver wire connected to a galvanic pile into his own rectum, an experiment the historian Stanley Finger calls "almost unimaginable."[35] This elicited all the unpleasant results Aldini had obtained with large animals, but performing the experiment on himself gave Humboldt the added benefit of firsthand experience. Thus he gained the insight that the involuntary fecal expulsions were accompanied by painful abdominal cramps and "visual sensations." Not content to stop there, he forced the wire further into his anus and found that "a bright light appears before both eyes." It would be hard to go further to prove one's dedication to understanding animal electricity.

In 1800, he undertook a journey to Venezuela for the purposes of investigating John Walsh's experiments with live eels, which didn't tend to survive the journey out of their native habitats. Using pack animals as bait to draw out the eels (some of which were five feet long and discharged 700-volt shocks—enough to stun the horses and mules) he saw for himself the unambiguous power of animal electricity. After the trip, he began to draw connections between this kind of powerful defensive biological electricity and the more quotidian variety that underpinned everyday motion and perception. In his subsequent writings on electric eels, he concluded, in carefully written prose, that at some future point "it will perhaps be found that, in most animals, every contraction of the muscular fiber is preceded by a discharge from the nerve into the muscle; and that the mere simple contact of heterogeneous substances is a source of life in all organized beings."[36]

Instead of adopting Aldini's strategy of charging ahead with his belief that Galvani had been right, Humboldt played the long game to bring experimental physiology back: he encouraged promising young scientists to study animal electricity. When Humboldt returned to Berlin from his travels in the late 1820s, he became a patron of the up-and-coming physiologist Johannes Müller and helped install him to lead the anatomy department of the world-beating university his brother Wilhelm von Humboldt had founded two decades earlier.[37]

The electroquacks had so thoroughly discredited the official record of animal electricity that when the first real evidence of its existence finally presented itself, even the scientist who rediscovered it didn't understand quite what he had found. In 1828, Leopoldo Nobili, a physicist in Florence, was working on ways to improve the sensitivity of electrometers, which were becoming increasingly important as they were crucial to the running of transatlantic telegraph cables. Electricians used them to confirm that current was flowing and that messages were being delivered. The early versions suffered from limited accuracy because the earth's magnetism interfered with the measurement of the wake of the current. No one could figure out how to get rid of it.

To do that, you needed a much more sensitive electrometer. (By now, these were starting to be known as galvanometers, thanks to a sly nod from the French physicist André-Marie Ampère.) To test that his improved version really was better, Nobili needed to find the weakest possible current. He remembered Volta's assertion that Galvani had witnessed not some special "animal electricity," but nothing more than the vanishingly faint currents generated from contact between two dissimilar materials. He realized that if his device could measure something as infinitesimal as the current traveling through a dead frog, its superiority would be incontestable. Sure enough, his new meter detected this flow, which he immediately christened "corrente di rana" (frog current).[38] It allowed him to make the first-ever recording of electrical activity from the neuromuscular preparation. But Nobili did not actually believe it was intrinsic to the frog—he was still firmly in Volta's camp. It was all, he insisted, to do with metals.

It would take another ten years for another scientist to correctly interpret the significance of what Nobili had measured, and finally put bioelectricity back on its pedestal.

The frog battery

Carlo Matteucci chopped the last of the frog thighs off its erstwhile owner and fitted it carefully onto the pile. He had killed ten frogs, removed their thighs, and carved each into a shape roughly resembling a halved orange—intact on one side, bisected on the other. He had then stacked these amphibian pieces atop each other into a biological inversion—some might say perversion—of a voltaic pile where zinc and copper were replaced with muscle and nerve. Matteucci had just finished constructing the world's first battery made exclusively of frogs.[39]

When he tested the current, he saw the output: the more thighs he connected, the more the galvanometer needle was deflected, indicating an increase in the flow of current. But that wasn't the end of the experiment. When he was satisfied that his pile contained enough biological material, he picked up the wire attached to his biological battery, and delicately touched it to a separate frog lying limp on a plate nearby—or rather, what was left of it. Unlike the frog battery, this one had been prepared in the style popularized by Galvani many years ago: flayed, head and midsection almost entirely missing save for the two crural nerves that still connected its spine to its legs, which remained whole. When the wire made contact, the hideous little half-puppet immediately jerked into the familiar dance. Animal electricity—and animal electricity alone—had caused a dead frog's legs to move.

Here, forty years after Galvani's death, was the first real progress in electrophysiology since the days of Galvani himself.

Matteucci was another of the promising young scientists who had been mentored and funded by Humboldt during the decades animal electricity fell out of favor. Humboldt had been inspired by Matteucci's enthusiasm for the promise of an underlying electrical force in the nerves, and had recommended the young scientist for the professorship he now held at the University of Pisa. He also defended Matteucci against attempts to discredit his discovery of the nerve centers torpedo fish used to control their shocks. So when Matteucci told Humboldt of his frog battery, his patron was so thrilled he immediately disseminated the paper to his entire social network, including Müller at the University of Berlin, who pressed Matteucci's paper into the hands of his eager young student, Emil du Bois-Reymond.[40] Humboldt was—again—a mentor to this young physiologist. "He is studying a matter, the deep natural secret of muscle movement," Humboldt wrote to the German cultural minister in 1849 to secure funding for du Bois-Reymond's research, "with which I, too, was preoccupied in the earlier half of my life." When he sparked du Bois-Reymond's interest with Matteucci's experiment, the match caught.

Though he found Matteucci's grotesque experiment unscientific ("no one can feel more deeply than myself how much this examination leaves to be desired in focus and clarity"), the work du Bois-Reymond did to build on it over the next two decades would finally resuscitate the long-dead field of bioelectricity and bring it back under the umbrella of legitimate scientific inquiry. He was incredibly ambitious and passionate about making his name, and du Bois-Reymond's fifty-five-year tenure at the University of Berlin turned into an attempt to secure his place in history by usurping Galvani as the father of animal electricity.

He was Galvani's heir in many ways. His commitment to good science and rigor was legendary. To characterize and measure the currents in nerves more precisely, he went to lengths that might be considered obsessive. He spent years on trial-and-error assembly of his own design for a special galvanometer sensitive enough to measure the faint electrical current flowing not through telegraph lines, but through frog muscles and nerves. He acquired so many frogs his Berlin flat turned into a "frog kennel."[41] When severing his frogs' muscle and nerve fibers, to avoid introducing any accidental source of external electricity, he would bite them in half instead of using a metal implement. He nearly blinded himself with his constant exposure to the irritants in frog skin. Berlin, like Italy decades before, began to run out of frogs. But this tenacity—inflamed by his determination to refine and take credit for Galvani's experiments—paid off.

Using his new galvanometer, du Bois-Reymond was able to see for himself the distinct electrical disturbance on his meter that accompanied muscle contractions. The needle on his galvanometer swung whenever the current passed through the area he measured. Whereas Galvani had only indirectly detected the electrical impulse traveling through the muscle by the evidence of a frog's leg twitch (making the frog the world's first galvanometer, in a twisted way), here du Bois-Reymond was directly seeing the animal electricity as it excited the muscle. The eighty-year-old Humboldt gamely lent himself as a guinea pig for these studies: even though he was now enough of a big deal to be taking "his daily dinner at the King's side," he would roll up his sleeve and flex until he deflected the needle of du Bois-Reymond's galvanometer.[42]

While most researchers greeted the early experiments coldly—the zeitgeist being still firmly against the idea of thoughts and intentions producing measurable electricity[43]—by the end of the eighteenth century du Bois-Reymond and his colleagues had successfully established bioelectricity as an aspect of neurobiology. The notion of electricity running through nerves and muscles was approaching respectability. A couple of outstanding questions remained: How did it travel? And why was this electricity so much slower than the kind that ran down telegraph wires?

But now you could measure it. Du Bois-Reymond and his colleague Hermann von Helmholtz called this electrical jolt that the nerve sent to activate the muscle the "action current." Other scientists soon joined the hunt to characterize it precisely, and while acrimonious fights broke out over many of the details, the existence of electrical phenomena in the nervous system became largely accepted. Du Bois-Reymond proved that electricity was relevant in the human body. Nerves ran on the stuff. He had made von Humboldt proud and usurped Galvani.[44] "I have succeeded in restoring to life in full reality that hundred-year-old dream of the physicist and physiologist, the identity of the nerve substance with electricity," he wrote.[45]

At the same time as du Bois-Reymond's research had resurrected the legitimacy of biological electricity, there had been new advances in mapping the brain and the nervous system. As had happened in the past, new tools cast doubt on old science, and a fresh uncertainty loomed. How could a single electrical impulse be responsible for such an enormous variety of discrete sensation and motion? Science at that point conceived of the nervous system as a vast uninterrupted network of fused strings. The best available metaphor was plumbing: Rather than being comprised of a bunch of separate cells, scientists still saw a series of tubes. Except instead of animal spirits, it was now electricity that coursed through them.

Thanks to better tools—like sensitive galvanometers and Volta's battery—and Humboldt, du Bois-Reymond, and Helmholtz's commitment to the rigors of the scientific method, the millennia-old mystery of animal spirits had finally been solved. Animal spirits, the things that carried the brain's impulse and intent to the limbs to carry out, and carried back the sensations of the world outside, were electric. Animal spirits *were* animal electricity. But instead of calling it animal electricity, the new term was "nervous conduction." It meant the same thing; it was just science instead of philosophy. Galvani was vindicated.

PART 2

Bioelectricity and the Electrome

"[A] full understanding of life will come only from unravelling its computational mechanisms."
Paul Davies, *The Demon in the Machine*

During the centuries of argument over the existence and nature of the nervous impulse, skeptics had many reasons to doubt that actual electricity runs in the animal nervous system. Investigations into the intriguing powers of electric fish and eels had yielded an obvious source: a giant electrical organ specialized to store up electrical charge and then dispense it in one big paralyzing zap. No anatomist had yet succeeded at locating anything like that in a human body. And without a power source, how were we supposed to be sending electrical current down our nerves? This led to lingering suspicions that electricity was just an unsatisfying metaphor for the otherwise mysterious conduction mechanism of nervous signals.[1]

All that changed in the latter years of the twentieth century—when we found the source. The new technique that aided this discovery caused a step change in the disciplines of electrophysiology and neuroscience. The consequent advances were so swift, and so numerous, that science historians Marco Bresadola and Marco Piccolino call them "comparable to that of quantum mechanics in Max Planck's day."[1]

CHAPTER 3

The electrome and the bioelectric code: How to understand our body's electrical language

By the end of the nineteenth century, animal spirits had been rescued from millennia of airy philosophical conjecture and placed onto the firm ground of the scientific method. Alexander von Humboldt, Emil du Bois-Reymond, and Hermann von Helmholtz had vindicated the work for which Galvani had given his life: what are the animal spirits in our nerves, these things that animate our every sense and motion? They are electric.

Yet even they could not have anticipated what their foundational tools and insights would set in motion over the next 150 years. Today, our understanding of bioelectricity is in the process of another metamorphosis as we begin to grasp the outlines of the electrome.[1]

The electrome transcends the bioelectric signals whose outlines Galvani and du Bois-Reymond glimpsed. Those were the drivers of the nervous system's ability to help us sense and move in the world, and today are well-characterized, thanks to copious investigations that helped establish the modern discipline of neuroscience. But in the past twenty years or so, a new picture has been emerging, one that shows with increasing clarity how instrumental bioelectric signals are beyond the nervous system, and how much they do in the rest of the body. Just as the genome describes all of the genetic material in an organism—the DNA that writes the instruction set to build it, the As, Cs, Ts, and Gs that make up the code this instruction set is written in, and other elements that control the activity of the genes—a complete accounting of our electrome would catalog all the profound ways different electrical signals shape biology.

Mapping the electrome would furnish a unique blueprint of the electrical properties that determine almost every aspect of our life and death. It would include a profile of the dimensions and properties of our electrics from the level of organs to cells to the tiny constituents of those cells, including mitochondria, and to the behaviors of the electrical molecules themselves.

As you saw in the first part of the book, the earliest glimpse of the electrome came to us courtesy of the electrical activity of nerves and muscles. "Animal spirits" became nervous conduction, and the scientific discipline that coalesced around its study was neurology. The insights from neurology (and electrophysiology, the field that united eighteenth-century electricians with theoretical physiologists) were codified in the 1960s into the formal discipline that is today known as neuroscience: the study of the animal nervous system.

The twentieth century brought tremendous advances in characterizing the patterns hidden in the electrical activity of the nervous system. We began to crack the code that explains how it transmits information to and from the brain. As you'll see in the next few chapters, almost all of these insights were delivered by probing the nervous system with metallic electricity. This led us to discover that artificial electricity could even, to varying degrees of success, modify our own bioelectricity—and health, thoughts, and behavior. That was startling enough, but then, toward the end of the century, we found out how much deeper the rabbit hole goes.

But before we go any further, we'll need to establish some of the basics of neuroscience, so that we can be on the same page about how the nervous system works and why people were so hot to prod it with artificial electricity. That's what this chapter is for. Please join me for a speedrun through 150 years of electrophysiology.

Nervous conduction 101

Figuring out how electrical messages are sent through the body got a lot easier once we figured out the structure of the brain and spinal cord and the special cells that enable them to pass communications. These are called nerve cells, or neurons. All this was established in a series of groundbreaking insights known as the Neuron Doctrine, which won Camillo Golgi and Santiago Ramón y Cajal a Nobel Prize in 1906. It was the first time we understood how the nervous system worked. (Before this, as the conversation around animal spirits makes clear, we thought the nervous system was just a single connected network of tubes that went from the brain through the whole body, which is why it made sense that you could fill them with water or hydraulic fluid—and why not much else about it made any sense.)

What Ramón y Cajal and Golgi figured out (though again via plenty of backbiting and disagreement) is that the nervous system was composed of cells—separate special cells that had been coined "neurons"—that could pass electrical signals from the brain to the nerves and muscles, and back.

No one had ever realized that the nervous system is made of cells, and that's because neurons don't look like your standard cell. Most cells look like a sphere that got a bit squashed. Not neurons.

A neuron has three distinct parts. It has a cell body (this bit does look like a normal cell), but sprouting from that cell body in all directions are branching protuberances of different lengths. These come in two flavors: the first—called dendrites—are very short, and they bring incoming messages to the cell body. The second—axons—can be up to a meter long, and their job is to send messages from the cell body to other neurons or muscles.

While some of the brain's 86 billion neurons exist only in the brain, a vast number of them extend down your spine and into your skin, heart, muscles, eyes, ears, nose, mouth, organs, intestine—in other words, into absolutely everything in your body—to make it move and feel and much more besides.

The "feeling" neurons that bring sensation and perception into the brain are part of the "afferent system," which brings you news of the outside world: the sights, sounds, smells, scratches, and bumps it confers on your body. These neurons are also called sensory neurons. The "moving" neurons, which bring your intention down into your body to actuate it, are part of the "efferent system," which lets you respond to the sensations carried by the afferent system.

Whether you're feeling or moving, the signals that are in charge of transmitting information to and from the brain are sent by way of an electrical mechanism: the action potential. This was the little needle-moving blip du Bois-Reymond knew as the action current or nervous impulse. Nerve impulse, action potential, you might have heard it called a spike—it's all the same thing: the small electric signal that relays a message between two neighboring nerve cells in the brain, or from nerve to muscle. When a dendrite receives a message, it passes the signal on to its cell body, which then evaluates whether to pass that signal on to the axon. If it passes the message on, it zips to the end of the axon, where it jumps to the next cell's dendrite. People had started fighting about whether the nerve signal was electrical or chemical almost from the moment du Bois-Reymond and Helmholtz started measuring it. But the fight graduated to near war with the discovery of how the signal hops from one cell to the next.

This is because that message hits a little speed bump at the end of the axon. There, it encounters a tiny gap that separates the axon of one cell from the dendrite of another. This gap is called a synapse, christened the same year the Neuron Doctrine won its progenitors the Nobel. The discovery of a gap between cells that were meant to be transmitting an electrical signal revived a lot of doubts about the still-fragile idea that animal electricity is real and that the nerve impulse is electric. After all, an electric signal can't travel over an air gap in telegraph wires, so why should it be able to do so in the wires of the nervous system?

In 1921, the discovery of chemicals called neurotransmitters, which float across the synapse's gap, only deepened those doubts. That briefly led to a fight over the nature of the nervous signal between opposing groups of scientists who called themselves the Soups (team chemistry) and those calling themselves the Sparks (team electricity).[2] It was like science's own *West Side Story*.

In the end—after fierce fights against the Soups' insistence that no electrical aspect existed—the Sparks won the day. Their champions were Alan Hodgkin and Andrew Huxley, two Cambridge University physiologists whose names may engender a very faint rustling in the back of your brain, in the part that had to memorize something about their endeavors in school. The reason their work is canon in the history of science is because they established that electricity is a crucial arbiter of the nerve impulse. They finally ended all the nattering between the Soups and the Sparks in the 1950s. Their experiments showed for the first time, in indisputable detail, exactly how the action potential is carried down a neuron by electrically charged particles, without whose electrical properties and activity nothing would happen at all.

Those particles are called ions. Ions are atoms with a positive or negative charge. The fluid that bathes every one of your cells—there's a lot of it, and it's the reason you always hear that you are 60 percent water—is teeming with them. The ions that are dissolved in this so-called extracellular fluid bear a strong resemblance to the constituents of seawater: mostly sodium and potassium, with smaller amounts of other stuff including calcium, magnesium, and chloride. Their precise concentrations inside and outside each neuron are the primary determinants of whether an electrical signal is permitted to pass.

They got their name from Michael Faraday, in honor of their weird tendency to move as if of their own accord. It was thanks to Volta's pile that he discovered this tendency, in fact. After Volta gave him one of his early prototype batteries in 1814,[3] Faraday used it to devise the principles of the electric motor and induction, and unify the laws of electricity. But much more important for our purposes here, it was instrumental to his discovery of the existence of ions. Faraday would experiment with putting various compounds into water and passing an electrical current through the water to see what it would do to them. Compounds are materials composed of a mix of two or more separate elements: under an electric current, the compound would dissolve back into the two separate elements that had made it, a bit like a cake neatly disgorging its sugar and flour. In this metaphor the "sugar" bits, having segregated themselves out of the mix, would migrate over to the electrode that was passing the current through the water; meanwhile, the "flour" particles would congregate at the other electrode.[4] Faraday didn't know what to make of this back then. What was it that was traveling through the water and accumulating on his electrodes? In 1834, he christened the mysterious particles "ions," which is about as far as anyone got with them for the next half century.

Then, in the 1880s, the Swedish scientist Svante Arrhenius realized that ions' movement was a result of them being pulled by electrical forces—which made sense, as ions were just atoms that, rather than being neutral, were either positively or negatively charged. This explained how they wandered through a solution as if of their own accord. They weren't doing that at all; rather, the positive ions had been attracted to the negative pole of the battery, and the negative ones wanted to get to the positive side. Finally, a clear explanation for Faraday's observations.

These properties apply in all solutions—including the biological soup that bathes the inside and outside of all the cells in all biological tissues. Ions are the ingredients that keep us alive. If you've ever been on an IV drip, you have ions to thank, and the nineteenth-century physiologist Sydney Ringer, who figured out the precise recipe of sodium, potassium, and other electrolytes to flood your vasculature with this facsimile of extracellular fluid. It enabled him to keep organs from failing even after they had been removed from the body that previously sustained them. His first experiment was on a frog heart, which, when placed into his new "physiological saline," was able to keep beating for several hours normally, sans frog.[5] The broth was originally called Ringer's solution, and has been monumentally consequential for biology.

But why were ions so important? What was so special about them that we couldn't live without them? As the twentieth century dawned, the consensus slowly emerged that ions could be the main agents in the electrical transmission of the nervous impulse.

As the Neuron Doctrine was coming into view, here's what we knew. One: biochemists had established that positively charged atoms like sodium, and negatively charged chloride, carried their electrical charge with them wherever they went. Two: it was also understood—thanks to people like Ringer—that ions populate the spaces inside and outside our cells. And finally, three: we knew action potentials created electrical activity strong enough to make galvanometer needles swing as the nervous signal flew past. Together, this was circumstantial proof that electric charges were moving in the nerve or muscle. But just as we had had a collection of unconnected facts about animal electricity in the eighteenth century with no framework to unite them, we didn't yet have a way to make sense of all these separate facts about the nervous system and ions. Not until the 1940s anyway, when a series of experiments showed exactly how ions were the main agents in the electrical transmission of the nervous impulse.

Alan Hodgkin and Andrew Huxley knew if they could prove that ion concentrations changed differently inside and outside a nerve cell during the course of an action potential, it would prove once and for all that electricity was involved in the very heart of the generation of the bioelectric signal—as a causative agent, not just as an echo of some chemical process.[6] Again, frogs met their end—but their nerves were definitely too small for any existing tool to be able to analyze the ion content inside their membranes. Next, Hodgkin and Huxley tried crabs. Still too small. Finally, they found an animal with a nerve big enough to stick an electrode into: the squid.

This creature's axon is uncommonly large—on the order of millimeters, a thousand times the diameter of the human variant, earning it the nickname "giant axon"—because it needs to send the brain's "run away!" instructions instantaneously through the squid's massive body.[7] This left plenty of room for Hodgkin and Huxley to insert the recording equipment necessary to monitor the electrical properties of the cell. They wanted to know how these would change as the nerve fired, and how the ion concentrations would change, inside and outside the cell, in response.

They found a way to stick one electrode inside, and another outside, and in so doing, for the first time measured the electrical difference between the inside and the outside of a cell. That difference was pretty big: the outside of the cell was 70 millivolts higher than inside when the nerve wasn't actively firing, just resting.

This number is called the cell's membrane potential. It measures the difference between the charged particles inside the membrane and outside it. Remember how ions are positively or negatively charged atoms? That means they bring their charge with them wherever they go. A sodium is carrying a +1, for example. So is potassium. Chloride drags around its -1, and I can't help thinking of it as being perpetually low-key ashamed. Fancy calcium poses conspicuously with its +2. Outside the neurons, a mix of these ions (and their various charges) congregates in the free space of the extracellular fluid. Because there's only so much room in any given neuron, the comparatively lower population of ions inside them creates a situation where the sum of charges is lower inside the cell than out. That's why it's 70 millivolts lower in any neuron than in the spaces outside it, and that 70 millivolts is exactly how the neuron likes it. For that reason, it's called the "resting potential." It's the neuron putting its feet up, conserving its energy.

But when an action potential raced through, Hodgkin and Huxley found those numbers changed wildly. The charge difference between the inside and outside of the cell quickly zeroed out, getting less and less pronounced until there was no difference between the inside and the outside of the cell. (Then, it kept going just a little bit past zero, until the inside of the cell was momentarily more positively charged than the soup outside.) But when all the fuss was done, it always returned to its 70-millivolt happy place.

While all this electrical commotion was underway, Hodgkin and Huxley also noticed that the different ions did very different things. During the resting potential, lots of potassium ions were inside the cell. But when the action potential happened, suddenly it was all sodium in the cell, which belched out a big wave of potassium. The cell's return to happiness was accompanied by a return of all the potassium ions. This phenomenon cascaded down the nerve, carrying the nerve impulse like a wave. That's how Hodgkin and Huxley finally proved that action potentials are unquestionably generated by changes in the concentration of ions.[8] The sodium and potassium were somehow responsible for the signal traveling down the axon—the electrical charge passed on by the precisely choreographed comings and goings of these ions.

So there it was: the solution to the mystery of Ringer's solution. The reason this precise mix of ions is crucial to keeping a body alive is that they are what makes the nerve impulse travel down the nerve. Without ions, the nervous signal couldn't jump. Then we couldn't breathe in or out or swallow and our heart wouldn't beat.

In 1952, Hodgkin and Huxley published the results of years of work that showed how sodium and potassium ions swap places in a cell, carrying their electrical charges in and out, to create the action potential. Revealing the mechanism of the action potential for the first time netted them the Nobel, but for Hodgkin the real triumph was the concrete evidence that electricity was not just a side effect but the cause. As he put it in his Nobel lecture in 1963, "the action potential is not just an electrical sign of the impulse, but is the causal agent in propagation."[9]

Their discovery was momentous, and should have set off a coordinated new quest to understand the information carried by these ions (and for a short time it did—one report quoted how the oceans around the main research centers briefly ran out of squid). But the spike in interest was short-lived. Just when animal electricity should have taken the spotlight once more, a cloud passed over the sun. No sooner had Hodgkin and Huxley revealed the elusive mechanism of the nerve impulse, than two other young researchers stole the show with a discovery deemed far more monumental: the double helix. In 1953, James Watson and Francis Crick—and Rosalind Franklin—unveiled their discovery of DNA. "There are only molecules. Everything else is sociology," Watson pronounced,[10] and the importance of bioelectricity was sidelined by a "bigger" discovery once again, just as it had been after Galvani.

Hodgkin and Huxley had shown that an action potential depends crucially on precisely timed coordination between sodium and potassium. But apart from the glamour of DNA, the bigger reason the avenue of research they had opened didn't continue into humans was because we didn't have small enough equipment to peer into the ever tinier nooks and crannies we'd have to explore to see ions going in and out of a cell and find out how it was happening. As a result, big questions went unanswered.

A long-standing theory going back to du Bois-Reymond's time had it that the cell's membrane wall would just disappear every so often, becoming transparent to a bunch of ions, like a curtain being pulled aside.[11] But that had never made much sense—and it made even less sense now in the wake of Hodgkin and Huxley's findings. Seeing sodium and potassium swap places like that made Hodgkin realize the membrane was not just being drawn back like a curtain. It was actively choosing what to let in and out. But what was the mechanism? Did neurons have special holes for particular ions?

How did nerve cells know how to only get rid of sodium, while potassium was untouched? This was especially weird given that potassium is about 16 percent smaller than sodium, lending extra mystery to the question of how a cell could momentarily expel all potassium while allowing in sodium.

During their long years of experimentation, Hodgkin and Huxley formulated a theory that the ions were entering and exiting through tiny holes that perforated the membrane like a sieve—maybe some of these holes liked sodium, and others potassium? People were starting to develop theories and language about these dynamics—but they didn't get a name until they were christened "ion channels."

I am ion man

What exactly is an ion channel, anyway? Since the 1960s, there had been growing suspicion that these pores were actually proteins that tunnel through the cell membrane. But no one managed to get any further until the early 1970s, when Erwin Neher, a physicist, and Bert Sakmann, a physiologist, got their hands on the problem at the Max Planck Institute for Biophysical Chemistry in Göttingen, West Germany. They reasoned that if those little holes actually existed, it should be possible to detect the teeny-tiny currents stirred up as the ions shuttled in and out. But since that's about a hundred thousand millionths of the current that powers your toaster, detecting it would require more sensitive equipment than anything that could be built.

So Neher and Sakmann created a new device that could isolate a little patch of neuron containing just a few, or maybe even a single one, of these putative holes. The ions and the holes were still too small to see with the equipment of the time, but when they were able to record the telltale current coming out of a single ion pore in a living cell membrane, Neher and Sakmann proved the holes were actually there. They existed.

What's more, they figured out how they worked. The shape of these current pulses made it clear that these little pores could only ever be in one of two states: all the way open or all the way closed.

They were never half ajar.[12] And when they were open, boy, were they open. A single open pore let potassium and sodium ions flood in and out of a cell at a rate ranging from 10,000 to 100,000 ions per millisecond. That is a lot of +1s.

A few years later, in 1978, William Agnew and his team at the California Institute of Technology (Caltech) finally identified what a sodium channel actually is: it's not just a hole in a sieve, it's a protein.[13] And with that insight, molecular biology went from show-stealing enemy to bioelectricity's best friend. That's because Watson and Crick's DNA discovery had given scientists the ability to read any protein's genetic code—which meant that if you could isolate and sequence it, you could clone it. And that meant you could start messing around with ion channels in earnest. You could make cells that only had shut versions, or open versions, and see what the effect was on an organism.

In 1986, Masaharu Noda was the first to clone a voltage-gated sodium channel (this is a kind of sodium channel that opens if it detects a change in the voltage pressing on the membrane around it).[14] Scientists started synthesizing proteins in different shapes and cloning different cells with different kinds and numbers of ion channels.[15] You could create cells with particular channels edited out altogether. If you were really enterprising, you might make cells with "designer" Frankenchannels that had been edited together—and see what happened next. This research soon gave scientists a full index of all the ion channels—sodium channels, calcium channels, chloride channels, potassium channels. Never mind the transparent curtain—it was these proteins that decided which ion was allowed to go where, when.

How did they make these complicated decisions? That puzzle was solved in 1991—the same year Neher and Sakmann got their Nobel for kick-starting this avalanche of research—by the biophysicist Roderick MacKinnon.

Many complex metaphors have been used to describe the incredibly complicated system MacKinnon uncovered. But I like to think of ion channels as shape sorters—you know, the toy you give a baby so it can shove different-shaped pegs into a wooden box through matching holes. Some of the pegs are round, others are triangular, or squares, or stars. The square holes accommodate the square pegs, and so on. So, while some holes may be *technically* bigger than their non-matching counterparts, they still won't fit through. They are incompatible with the dimensions of the channel, and therefore it remains impenetrable. (It's actually a little more complicated still, because the holes in this baby toy shape-shift to accommodate the peg they like best.)

After MacKinnon completed the picture of the cell membrane, we understood for the first time the array of interlocking mechanisms that underpin biological electricity: how the proteins in the membrane work with ions to generate the action potential, and how everything goes back to square one again when the action potential has passed. Once we understood ion channels, we were able to understand the action potential in full.

It is remarkably similar to how you manage an exclusive night club.

In the club

I should mention that the following analogy ignores the entire universe of complexities inside and outside a cell, focusing only on the place where the voltage is generated. But after all, this is a book about bioelectricity.

You can therefore think of the cell as a highly micromanaged nightclub. Ions take the role of the patrons, and the ion channels act as bouncers staffing VIP access doors. These participants orchestrate the three stages of the action potential. (My ridiculous ion nightclub piggybacks on a wonderful explanation in Frances Ashcroft's *The Spark of Life*.)

Stage 1—the resting potential

At rest, when no action potentials are passing through, the nerve cell is in what's known as its "resting potential." This was the 70-millivolt difference Hodgkin and Huxley found. In this state, the inside of the cell is more negatively charged than the soup in the extracellular space.

Inside our club, the crowd is mostly made up of positively charged potassium ions, which cram into the tiny space at a concentration fifty times greater than they mill around outside. Outside the club, a long line of hopefuls—mainly sodium ions, also carrying a positive charge—press against the doors. But alas, most of those doors are shut to them. Management has a clear preference for members of the potassium group—a strict NO SODIUM ALLOWED policy is in effect. Those sodium ions though, they're just like us, so they cram in greater and greater numbers against the doors, wanting to get in. But management is not messing around. If a bouncer, known as an ion pump, finds a sodium ion has somehow sneaked in, it is briskly escorted out—and in a final insult, any three available potassium ions are waved through security in its place.

As for the potassium ions—well, they're just like us too. They get tired of the heaving conditions inside the club and occasionally depart, leaving a negative charge behind them. There are no barriers to them leaving.

This delicate balancing act of managerial vigilance, the sodium ions' desperation, the potassium ions' general standoffishness, and other variables, is what keeps the cell membrane's resting potential perpetually hovering at -70 millivolts. (More positively charged sodium ions outside than positively charged potassium ions inside make the inside negative compared to the outside.)

Little wonder the cell biologist Robert Campenot describes the state of a nerve cell before an action potential as being on a hair trigger. All it needs is an excuse. Any minute change in the equilibrium will pitch it all into chaos.

Before we get to that, though, there's just one more thing. Those doors with the bouncers I've described so far are not the only doors. There are also emergency doors: the voltage-gated sodium channels. They bang open if they sense a change to the carefully maintained resting potential. If the amount of charge outside the club changes the right amount, then in an instant, all the potential energy that has thus far been held in delicate suspension is released.[16] Which is to say, if the crowd of sodium outside the club gets too rowdy, the velvet ropes topple, and they start to force themselves into the club. And then it's panic at the disco.

Stage 2—the action potential

Relative to the size of the cell, the membrane potential is massive. The cell membrane is about 10 nanometers across, and one side of it is 70 millivolts more negative than the other. If you had the equivalent voltage difference across your body, it would feel like 10 million volts. That is . . . a lot. An intense static shock that will make you turn the air blue with swear words packs about 10,000 volts.

The shock of an action potential is way more intense for our little ion channel bouncer friends. The emergency doors are flung open. The sodium ions take advantage of the confusion to rush into the club, triggering a feedback loop: the bigger the change in membrane potential, the more sodium channels open and the more sodium ions crowd in—and the more sodium ions come in, the more the voltage goes positive, the more sodium channels open, and so on. In an instant, sodium has taken the place by storm.

Step 3—repolarization

Now that millions of sodium ions have mobbed this formerly exclusive club, jostling and crowding the horrified potassium ions, the inside of the cell can briefly reach 100 millivolts more positive than the outside. Less than a millisecond later, the potassium channels open and the potassium ions, disgusted, leave the club en masse.

Inside the club, the mass exodus of potassium ions has returned the cell membrane to its resting state. But now the customers are all wrong! Management is desperate to win back the fleeing potassium ions. They lock the place down again. The bouncers start cracking their knuckles. Most of the sodium ions leave of their own accord.[17] It takes a while to convince the potassium ions to re-enter the despoiled premises. Eventually, though, management coaxes them back. Then it's just a matter of time before the whole thing happens all over again.

The voltage opens and closes the channels.[18] Sodium and potassium channels that react to the voltage changes mediate the generation of action potentials, which allow signals to be propagated from one end of a neuron to another. The same mechanism was ultimately even what controlled the chemical neurotransmitters. At the very end of the axon, where the action potential terminates, there's another set of bouncers: voltage-gated calcium channels.

When the action potential hits, these open, and calcium from the extracellular saltwater floods into the end of the axon. That releases neurotransmitters (your familiar serotonins, dopamines, oxytocins) to drift across the axon terminal and onto the nearby entry point of the dendrite of the neuron next door. And that sets off the next action potential, and the whole sequence starts all over again. All these things—chemical and electrical aspects—are ultimately controlled by the membrane's voltage, its electrical status.

And that's the story of the nerve impulse—the thing that's responsible for our every sensation, motion, emotion, and heartbeat. Its electricity is the central generator. The source of your electricity, and mine, is not a separate electric organ like the one in an eel, but a self-regenerating mechanism produced inside the cells themselves, by the exquisitely coordinated dance of ions through proteins.

The basic mechanism that was in charge of all of this complexity was startlingly simple. Stack more charged ions on one side of a membrane than another, and you got an electrical potential. Change the voltage, and you released all that energy. That's essentially how a battery works: one side has a different amount of charge than the other. Nerve and muscle cells, it became clear, were tiny rechargeable batteries.

Forty trillion batteries

But they were not the only cells that acted like batteries. After it became possible to properly study ion channels using the tools of molecular biology, it became clear that ion channels (and the ions they admitted and didn't) were also present in every other cell in the body. This was the wake-up call: what were they doing there?

What use did all these other cells have for electrical properties?

We found out in due course. In 1984, the ion channel physiologist Frances Ashcroft found that the pancreas, for example, uses a particular potassium ion channel to issue the electrical commands that precisely synchronize its insulin-secreting beta cells. (These travel ten times faster than the chemical kind, so it's the only way to synchronize that many cells to act in unison.) This potassium channel needs to be in perfect working order to coordinate the insulin release. In the early 2000s, Ashcroft and Andrew Hattersley found the mutation that locked it open—and thereby caused a variant of diabetes.

Insights such as these multiplied and soon transformed medicine. Ion channel physics emerged as a major biomedical discipline in its own right. Now scientists had the tools and understanding to investigate how ion channels in muscle and nerve cells sustained the most basic workings of the human body. And more importantly, what happened when they didn't. Most importantly, they finally had a new tool in the arsenal for manipulating that electricity more precisely, and it would prove the most consequential tool for bioelectricity researchers since the invention of the battery.

The first ideas for drugs that could manipulate the electrics came from neurotoxins. In the 1960s, research on neurotoxins had clarified that many of these natural poisons affect sodium and potassium balances, throwing into chaos the delicate mechanisms that make cellular communication possible—acting like a reverse Ringer's solution.[19] The reason you're not supposed to eat puffer fish (unless it has been prepared by someone with an advanced degree in precision-filleting a puffer fish) is that some of its parts carry a defensive poison called tetrodotoxin. Get even the tiniest amount of that inside you, and it can quickly paralyze the muscles that drive everything in your body, including your lungs, and then you asphyxiate. The exact mechanism of how this works was unlocked by the improved understanding of ion channels brought by Neher and Sakmann: tetrodotoxin keeps sodium ions from entering the cell.[20] It wedges itself into those channels, blocking the doors, and if there's no sodium influx, then there's no outward potassium stampede, and that prevents all the rest of the dominoes that cascade into an action potential. Other kinds of neurotoxins pry open all the doors, which has the same eventual effect: the cell can't communicate any signals to other nerves or muscles. No cell can survive without functional ion channels.

Once researchers realized how nature made neurotoxins—by throwing a wrench into those all-important ion channels—they realized they could make their own bespoke neurotoxins to jam shut, or pry open, only the channels of their choice. (Ashcroft and Hattersley figured out that an existing drug could close the errant ion channels and reverse this rare form of diabetes.) And that was the dawn of the age of ion-channel drugs.

Ion channel drugs are a bedrock of modern medicine. They underpin treatments you get for some snake bites by artificially pumping up the communication between the nerves and muscles. They underpin drugs for heart arrhythmias. Now researchers are investigating a whole slew of movement disorders, epilepsy, migraines, and some rare inherited diseases for possible mutant ion channels.[21] All over biology, ion-channel physics revolutionized the treatment and conceptualization of disease and disorder. "It is difficult to exaggerate our misunderstanding of heart action potentials before we knew about calcium channels," wrote one cardiac electrophysiologist.[22]

Ion channels are important drug targets, but our understanding of them is incomplete. We keep finding more unexpected variations of them. One is gap junctions, which were first noted in the heart but now seem to be in every one of our trillion cells. A gap junction is a special ion channel that pokes between two neighboring cells, creating a sneaky door only they share, like adjoining hotel rooms. In heart cells, gap junctions synchronize the activity of cells that need to operate in tandem, but they also festoon the membranes of skin cells, bone cells, heart cells, and they even occur on blood cells. They are everywhere. They all talk to each other using these electrical synapses. What on earth for?

New ion channels aren't the only surprises. Another recent observation is the electron current expelled by cancer cells as they make the transition out of good health.[23] On a larger scale, there are also aspects of the nervous system we didn't appreciate until the turn of the twenty-first century, when it started to emerge that the nervous system doesn't just act on the feeling and moving bits, but also regulates organ function and the immune system. These are the kind of insights that are beginning to form the outlines of the electrome.

Until recently, knowledge of these disparate electrical features of biology was sequestered in narrow subdisciplines. That's because the study of bioelectricity had been increasingly siloed into neuroscience, and in electrophysiology, which focused a lot on nerves and neuroscience—to the extent that scientists assumed bioelectrics were only used by nerves.

One of the more astonishing features of the electrome is that animal electricity is by no means confined to animals. We're not the only ones with these ion channels. All the other kingdoms are run on the same stuff.

Electric kingdoms

We had had glimpses of that reality, too, much earlier than we could reasonably account for it. In 1947, the physiologist Elmer Lund found electric fields coming off algae.[24] He wasn't alone;

these confounding electric emanations seeped from every other biological surface people thought to measure: Venus flytraps, frog and human skin, fungi, bacteria, chick embryos, fish eggs, and oat seedlings.

Reports from disparate fields of study indicate that the electrical signals used by plants, bacteria, and fungi are all weirdly similar to our own, and the research is beginning to suggest that they use these signals to very similar effect. Bacteria use electrical potassium waves to coordinate themselves into biofilm communities (disrupting these electrical control signals is a hot research topic in the fight against antibiotic resistance).[25] Fungi use calcium to communicate along their long tendrils whether they've found a nourishing food source or a dud.[26] Plants use electricity to activate chemical defenses against predators. The list goes on and on.

We have wondered in the past twenty years, as we discover ever more similarities between their electrics and ours, why these signals (in bacteria, in fungi, in protists) are so similar to those in our nervous system. But now a lot of people are starting to wonder if perhaps we've been getting the question backward: why are we so similar to *them*, and what does that mean about our electrics?

All creatures, brains or no, use a collection of similar ions to create voltages across their cells. We all use these voltages as a basis of communication. Animals use them to make their nervous systems function as a command and control center; other kingdoms use them for signaling and communications without a nervous system. "Flipping the voltage potential is how, I think, all signalling probably began," says Scott Hansen, an electrophysiologist at the UF Scripps Biomedical Research Institute at the University of Florida.

And that is raising a wild idea: could we have another communications system running in parallel to the nervous system? Recent research strongly suggests our bodies are running at least two—if not more—electrical communications networks.

Evidence has begun to accumulate that the bioelectricity in the nervous system—the animating force behind animal spirits—is not the only electrical communication network used by the animal body. Strange electrical features and behaviors connect all the cells in our body. Skin, bone, blood, nerve—any biological cell—put it in a petri dish and apply an electric field, and they all crawl to the same end of it. It's as if they can sense the electric field, even though we don't yet understand how cells could possibly sense those things. All we know is that electric fields affect the bioelectric properties of a cell—any cell, and sometimes whole organs—in a way that can be used to make it do things it normally wouldn't.

It is for this reason that some scientists are beginning to think bioelectricity can be understood as a component of epigenetics—which describes how the environment can cause changes that alter the way your genes work without changing the actual DNA. "More and more epigenetic factors which drive the organization of biological information patterns and flows are being discovered," writes the physicist Paul Davies.[27] Bioelectricity is emerging as a major—if as yet poorly understood—epigenetic factor, he thinks, providing a powerful way for cells to manage epigenetic information. But other researchers are finding that it may be more than just another aspect of epigenetics. The word "epigenetic" means "above the genes." And maybe electrical signaling functions as a kind of "meta-epigenetics"—one ring to bind them, if you will. As you will see over the next few chapters, electrical guidance exerts control over a great many complicated aspects of biology, from how genes are expressed to whether inflammation will commence in the immune system.

The bioelectric code

A granular understanding of the electrome, then, could also provide a way to control the genome almost as easily as we can control our computer hardware with software. Indeed, the Tufts University researcher Michael Levin is among those who have found evidence to suggest that the electrical dimensions of life can exert control over genes, providing a way to hack other systems we previously thought were too complex to precisely control. Levin suspects that this deeper understanding of bioelectricity will yield a bioelectric code. This code is written not in genes but in ions and ion channels. That code controls the complicated biological processes that formed you in the womb, by executing a controlled program of cell growth and death. The bioelectric code is the reason you retain that same shape throughout your entire life; it prunes your dividing cells so you keep being recognizably you. And if it could be deciphered and manipulated, it could be used to precisely re-engineer the human physical form, rescuing it from birth defects and cancer (more on that in Chapters 7 and 8). If we can profile the electrical properties of biological tissues in the same way we have profiled its genetic basis—that is, to complete the human "electrome"— we can crack the human bioelectric code.

When people think about the origin of life, obviously the first thing that comes to mind is the genetic code. How did DNA and RNA evolve in the first place, and lead to reproducible life, and all that jazz? There's a second thing that should come to mind, but usually doesn't. How is it that you got a cell membrane?

The cell membrane is important for a number of reasons. The first is just practical. All the DNA and RNA in the world, reproducing all the elements, all the nucleotides and amino acids you could possibly want for life—well, they'd just float away in a big soup if they didn't have a container. To do anything remotely useful with the constituents of life, you need something to hold them all together. So that was the membrane: the most underappreciated evolutionary innovation.

But there's a bigger reason the membrane is so important. As soon as there's a membrane, there's a separation between an inside and an outside. And since every cell we know about has always contained different kinds of ions, the second you had a membrane separation, you had a voltage. That's just physics. After that, you just needed the proteins to form passages in the membrane that allowed all those ions to get in and out of the cell.

These ion channels, as a group, are something like 3 billion years old. Plants, fungi, animals, the whole lot of us inherited them from our eukaryotic ancestors. Signaling certainly didn't start with sodium channels—those only evolved around the time the first nervous systems did, about 600 million years ago.[28] In 2015, the neurobiologist Harold Zakon published a deep evolutionary history of ion channels and found most of the same ion channel families are present all the way back to our last known ancestor.[29] Our sodium channel's building blocks, Zakon found, were found in the first ion channel, potassium. In fact, the potassium channel is the little Lego from which most of the other channels—sodium, calcium, and so on—were later formed. "The motif that allows potassium to permeate the channel is very ancient, very conserved. It's pretty much the same from bacteria to us," Zakon says. "We have it, every cell in our body has it, probably every cell on earth has that gene for that channel."

In fact, you can still find that molecular motif of the first ion channels in bacteria today. Every subsequent channel and pump comes from that ancestral gene.

The upshot is this: separating and moving ions across membranes is fundamental to all living things. Nervous systems did not invent it, and we are still nowhere near understanding the full extent of how nature recruits its electric potential. Although literally all cell types use this self-generated electricity, the breathtaking range of functions for which it is instrumental is utterly underappreciated. It's certainly not covered in textbooks of introductory biology, at least not in such a way that would make anyone actually appreciate the importance of the electrical dimension of life or the deeper significance. Those elements we all ferry across our membranes—sodium, calcium, chloride—are fossil stardust. If there are any other cells out there in the universe, we might share that with them, too. "Probably every cell in the universe," Zakon noted.

We knew none of this when we first started to experiment with animal spirits and found the first hints of what would later become the bioelectric code. We didn't know about ion channels or patterns, and the only tools we had to probe the animal spirits were cousins of Volta's pile. Which is why the first glimpse of the electrome came to us courtesy of the electrical activity of nerves and muscles. This—as you'll see in the next three chapters—was how we started to learn that we could use electricity to take control of our hearts, brains, and central nervous system.

PART 3

Bioelectricity in the Brain and Body

> *"Wonderful as are the laws and phenomena of electricity when made evident to us in inorganic or dead matter, their interest can bear scarcely any comparison with that which attaches to the same force when connected with the nervous system and with life."*
> Michael Faraday, *Experimental Researches in Electricity*

In the twentieth century, better tools began to reveal the first hints that there were patterns in bioelectric signals that might indicate health or disease. This quickly led to the notion that electrical stimulation could be used not only to understand the body but to improve it—by overriding the faulty patterns with healthy ones. We could electrically control ourselves back to health.

CHAPTER 4

Electrifying the heart: How we found useful patterns in our electrical signals

Few protests had been lodged about the horrific dissection of either Galvani's frogs or Aldini's decapitated prisoners in the quest to understand animal electricity, but the dog-loving citizens of the UK had their limits. In 1909, an affronted member of the anti-vivisection lobby arrived at the House of Commons with a report of an alarming act of scientific cruelty.[1]

That May, the lobbyist had attended a "Conversazione"—a soirée at which the scientists of the Royal Society would demonstrate their findings for members of the public. (The allure of these events, according to one newspaper, was that "for once [scientists] condescend to admit the average man and woman into their mysteries.") One such demonstration featured a scene shocking enough to warrant a hearing before Parliament: a dog had been restrained by a "leather strap with sharp nails wound around his neck," according to the anti-vivisectionist's complaint, ostensibly to keep the poor creature immobilized while its "feet were immersed in glass jars containing salts in solution, and the jars in turn were connected with wires to galvanometers. Such a cruel procedure should surely be dealt with under the Cruelty to Animals Act of 1876?" the petitioner admonished.[2]

The grisly description proved somewhat misleading, and it fell to acting Home Secretary Herbert Gladstone to clear things up.[3] Rather than an ill-fated specimen destined for experiments, he explained that the animal in question was in fact the scientist's beloved pet English bulldog, Jimmy. That "leather strap with sharp nails"? Jimmy's (rather expensive) brass-studded collar. And finally, Gladstone clarified that the "solution" in which the dog stood—of his own volition and indeed pretty cheerfully, in keeping with his reputed "Churchillian" demeanor—was saltwater. "If my honorable friend had ever paddled in the sea, he will appreciate fully the sensation obtained thereby from this simple pleasurable experience," he concluded. Nonetheless, with this harmless demonstration Jimmy the bulldog had just done more for the advancement of electrophysiology than all of Aldini's dead prisoners. He—or rather his owner, the physiologist Augustus Waller—had demonstrated the world's first recording of the electrical activity of the heart.[4]

The ability to listen to electrical signals would soon become a cornerstone of modern medicine, and not just for the heart, whose previously opaque processes were about to become a lot more transparent. By the end of the twentieth century such signals would be discovered radiating from many other organs, using tools Waller wouldn't have dreamed of, to gain penetrating insights into the health and disease of a person's body and mind to a degree he would hardly have believed possible.

The telltale heart

In the mid-1880s, Waller realized that if you connected your limbs to an electrometer, it should be possible to form a circuit through which you could conduct the heart's electrical signal and make it legible. (Prior to his breakthrough, the only way to "read" a heartbeat had been to open the body and put electrodes directly on the exposed organ—a feat only possible with more grisly animal experiments and occasionally on people with horrible, yet medically fortuitous, injuries.)

For Waller, however, recording the electrical activity of the heart remained something of a party trick. The tracings provided by his set-up were fuzzy and imprecise, owing to the slow response time of the equipment.[5] They couldn't tell you much about the heartbeat except that it was there. Indeed, this is how his guests tended to avail themselves of it at his soirées, where the ladies and gentlemen in attendance used his contraption to bring solid evidence to their companions that they were in possession of a beating heart. And what a contraption it was: the cumbersome set-up required polite company, after dinner, to remove one shoe and sock, sit in a chair connected to a large measuring instrument called a capillary galvanometer that looked not unlike a Victrola cabinet, and dip a bare foot and one hand into two buckets of saline water. If the unusual set-up made them a bit nervous, Waller would offer to first demonstrate on Jimmy, who placidly endured the whole thing.

Instead, the Dutch physiologist Willem Einthoven saw the potential that escaped Waller's notice. In 1889, at a physiology conference in Switzerland, Einthoven witnessed a demonstration of the technique by Waller himself. He soon refined the device to do what Waller's couldn't: get tracings precise enough to read the contours of the signal.[6] Over the next decade, steady technological improvements led to ever more exact recording of the heartbeat, culminating in 1901 with Einthoven's particular contribution, the "string galvanometer." This electrometer was capable of measuring the faintest of the body's electrical signals. If you'll permit a gross oversimplification, its mechanism was a string, illuminated by an extremely bright light to cast an exaggerated, enlarged shadow onto a white sheet. You could watch that shadow vibrate with every beat of a heart. Einthoven refined it further by including silver-coated quartz strings, moving photographic plates, and mechanical pen recorders, but my description of the basic mechanism stands.

The only reason Waller or Einthoven were able to pick up these surface readings from the heart is that the signals, as small as they are, are so incredibly "loud" in combination—loud enough to be picked up by that string galvanometer. An individual cardiac muscle firing its action potential is a friend humming quietly next to you; loads of them firing in synchrony is a 100-person choir harmonizing with the organ in the last four glorious chords of Handel's *Messiah*. Only a few places in the body perform Handel's *Messiah*: many heart muscle fibers have to fire together simultaneously to produce the heart contractions that pump blood around our bodies.

Where Waller's tracings had been fuzzy and imprecise owing to the slow response time of the old and busted type of galvanometer he was using, Einthoven's better version yielded sawtooth waveforms of such crisp resolution that it could even tell a healthy heart from a sick one. It was Einthoven who put a name to these squiggles at the Dutch Medical Meeting of 1893, where he coined the term "electrocardiogram," now best known by its acronym—ECG.[7]

The machine he had built to do it, however, was a monster. Waller's early iteration was dwarfed by the sheer size and ungainliness of Einthoven's creation, which filled two rooms, weighed 600 pounds, and required five human operators and special cooling equipment[8]—not to mention that the person whose heart was being read now had to immerse *two* hands along with the usual foot. But the equipment worked a treat: in the early years of the twentieth century, Einthoven further formalized Waller's fuzzy scribbles into the diagnostically accurate troughs and peaks, within whose characteristic idiosyncrasies doctors could diagnose heart conditions in hospitals. Clinicians began to acquire them, including the cardiac electrophysiologist Thomas Lewis, who in 1908 started using it on his patients at University College Hospital. With this newfound ability to investigate and describe various heart-rhythm abnormalities, beginning with atrial fibrillation, Lewis knew he was laying the foundation for a new field: clinical electrocardiography. The electrocardiograph enabled medicine to peer into the body as never before, and over the next decades helped explain exactly how the heart's electrical activity was instrumental to its ability to coordinate the flow of blood around the body.

A pump controlled by electricity

Each pump of blood through the heart is set into motion by a group of cells that is best understood as the conductor. The group of cells, located in the upper right part of the heart, is called the sinus node. This conductor coordinates all the cells of the heart into a precise rhythm that ensures blood only ever enters one specific kind of chamber and only ever exits another specific type. Blood enters a set of upper chambers called the atria, and flows down to the ventricles (the lower chambers), which contract about half a second later, one ventricle sending its blood to the lungs and the other sending its blood around the body. This is quite a precise rhythm to orchestrate, and high stakes! If you do it wrong, the heart can't coordinate the blood distribution around the body properly and the body will die. And it all depends on the electrics.

The conductor kicks off all of this with an action potential, but these are not the familiar action potentials of the nervous system. That's because the muscles of the heart don't have their own nerves driving them in the same way that nerves drive skeletal muscle. The heart is all muscle, but it's an unusual kind of muscle. It's a kind of self-determining muscle that moves without you being in charge of it—as you well know, your heartbeat is not under your control. With a lot of practice and focus, you can learn to slow it, but you can't stop it in the same way you can close your eyes. Rather like nerves, the heart muscles generate their own action potentials, except without chemical synapses.

So how does the action potential pass from cell to cell? How is the conductor's signal sent to all the muscle cells through the heart? Turns out, instead of being connected by standard synapses, they're all connected by direct, electrical high-speed lines—those gap junctions I referred to in the previous chapter.[9] These adjoining hotel room doors are usually left open so the signal can instantaneously zap between rooms. Whatever one cell knows or experiences diffuses immediately through the connecting door for its neighbour to know or experience instantaneously. This mode of communication is about ten times faster than a regular chemical synapse because it does away with neurotransmitters and gaps between cells.

This is how the heartbeat's rhythm shimmies down from the top of the organ to the bottom to ensure that the outgoing blood is always pumped just precisely half a second later than the incoming blood comes in.

It was this synchronized wobble that Waller was picking up. But his early equipment was too primitive to see the details, which were only made legible when Einthoven deployed his fancy string. That's when we got our first look at those sawtooth-shaped blips that you may know from medical dramas (or if you've ever been hooked up to a heart monitor).

What was much more interesting than seeing the normal heartbeat, however, was that Einthoven's crisper readings made it possible to see when things were *not* all right with the heartbeat. You could now not only *visually* distinguish the signature of a healthy heart from a sick one, you could start to detect specific ailments—for example, an abnormally slow heartbeat. This condition, called bradycardia, means the blood can't deliver enough oxygen to the brain and other body tissues, so a person with this condition often feels dizzy or weak, or faints.

Long before we fully understood how all these signals traveled and worked, people started to use electricity to whip those errant signals into shape.

The pacemaker takes control

The pacemaker had its origin on an operating table in Prussia in 1878. Catharina Serafin had just survived a brutal surgery that had removed a malignant growth but left her beating heart exposed, covered by nothing more than a thin flap of skin.[10] This afforded a rare opportunity for the German physician Hugo von Ziemssen to mechanically and electrically stimulate her living heart, which led to the new realization that it was possible to directly act on the heart with electricity. Previous investigators like Aldini had thought the only way to manipulate the heart electrically was by way of the nervous system.

In the course of experimenting on Serafin's heart, Ziemssen realized that if you applied periodic pulses of DC current—the same steady flow of electricity Volta had generated from his pile—that was just a bit faster than the natural heart rate, the heart would try to keep pace with this artificial metronome. Here was evidence that you could overwrite a faulty rhythm—or resuscitate a stalled one—by dubbing an artificial electrical pulse into the place at the top of the heart where the natural electrical signal originates. But nothing much came of it. This method only worked when you put the electrode directly onto the exposed surface of the heart—it didn't work if you applied the pulse through a closed chest. And with no one clamoring to open their body to apply electric shocks, there was no business case.

It took thirty more years for any new medical applications to come of this understanding. What finally got things going was that the electrification of America had led to a sharp uptick in accidental electrocutions[11]—exactly the kinds of "temporary deaths" Aldini had been trying to find a way to reverse more than a century earlier. Now the matter had some urgency. It had been established that you could restart or correct a heartbeat—the next question was how to keep it going. Work got underway on a device that could overwrite the heartbeat continuously. And that device was absolutely terrifying.

It was the size of a small suitcase, weighed 7.2 kilograms—about 16 pounds—and was operated by a hand crank.[12] A wire sent the electricity it generated to a needle that pierced the heart. It worked, but it was a tough sell to get this into clinical trials. Finding the exact right spot to put the needle was of paramount importance; if you got it wrong you got a fatal hemorrhage. In 1932, both the device and its creator Albert Hyman were roundly condemned by the American Medical Association: reports of cardiac resuscitation by such cardiac injections, they stated, "belong with miracles."[13] The skepticism was part of a hangover from years of experiments like Aldini's, and ensured that no American manufacturers were willing to put their reputations on the line to help Hyman produce his device.

Nonetheless, by 1950 other physicians, clearly in need and now able to avail themselves of more advanced materials, developed a different design. These were not always what you might call an improvement. People had to wheel them and their tangle of tentacular cables around on a trolley. They sometimes required power from the wall (more's the pity if the power went out, which was not unheard of). Early attempts to figure out how to make them implantable—the ultimate in portability—needed a better power source than what was available.

Lightning to the heart

If you think nuclear power would be a bad choice for powering an implant close to your heart, there are 139 people who would disagree with you.[14] In the 1970s, several manufacturers rolled out pacemaker designs that were powered by a bit of plutonium. The heat generated as this radioactive isotope decayed was transformed into electricity that powered the circuits in the model,[15] but don't worry, they were "shielded well enough to deliver very little radioactivity to the patient." Battery designs for implantable pacemakers only got weirder from there, including one that ran on biological electricity not conceptually dissimilar to Matteucci's frog thigh battery.[16]

In 1958, Wilson Greatbatch found an enduring source of power that was less unsettling than plutonium, with the invention of a pacemaker that used a lithium ion battery, and there we still are today, for the most part.[17] Within a couple of decades, his invention had been refined into the small, implantable gadget we understand today as the pacemaker.

The concept is simple enough and the pacemaker is implanted much the way Hyman did it. Fortunately, no one uses a needle to pierce the heart anymore. Instead, an electrode is surgically implanted on the faulty spot that's causing the trouble. The electrode is attached to the pulse generator by a wire that carries the stimulating electrical charge, conceptually not entirely unlike the kite string Ben Franklin used to bring lightning down from the sky. Except instead of atmospheric lightning, this conductor conducts the tiny, contained lightning from a battery-powered stimulation unit: the pacemaker. These are improbably tiny now, considering their trolley-top origins—about the size of a ten-pence piece—and they keep getting smaller.

The most common use for a pacemaker is to speed up a slow heart rate (as in bradycardia). The tiny lightning overwrites the heart's own bioelectricity, applying miniscule, regular electric shocks to drive the heart at the right rate.

When it reaches the muscle cells in the sinus node (that conductor in charge of the first domino), the electrical stimulation forcibly changes the membrane potential of the cell.[18] The muscle is depolarized, which opens the voltage-gated sodium channel, and then triggers the action potential. And then that sets off all the rest of the cascading actions of the heartbeat.

Some of the most advanced models today don't just zap; they can listen to make sure they're dispensing the right kind of zap at the right time. They sense the wearer's heart rhythm so as to modulate it in real time. That ability to respond to real-time feedback puts the pacemaker in a category known as closed-loop devices.

After Greatbatch added the lithium ion battery, things happened fast. By the 1960s, several of the twentieth century's biggest technological breakthroughs—plastics, transistors, microchips, batteries—conspired to make the pacemaker implantable and reliable.[19] Engineers and scientists who adapted it into a working device went on to found a medical device company called Medtronic. Over the next twenty years, the number of patients with pacemakers quickly rose from half a dozen to nearly half a million.

ELECTRIFYING THE HEART

In the late 1960s, a neurosurgeon in Wisconsin took Medtronic's implantable cardiac pacemaker outside its intended environment for the first time, repurposing it for his chronic pain patients. It was implanted in the spine—but that was just the beginning for the pacemaker's strange journey, which would soon find a new home in the brain.

So did Waller's early tracings, for that matter. After first allowing doctors to diagnose heart conditions in hospitals, and then laying the groundwork for the first recordings of the activity of the brain, the electrocardiograph became the basis of much of the electrical imaging that proliferates today to diagnose sleep and neurological disorders. These advanced brain diagnostics, in turn, opened the door to the idea that animal electricity is the body's way of digitizing information so that it can speak to itself in a kind of specialized neural code, an idea that took root in the twentieth century and has flowered into the defining idea of neuroscience in the twenty-first. Many are now convinced that, with these descendants of Waller's early contraption, we are mere steps from reading the electrical activity of thoughts—and perhaps unlocking the secrets of consciousness itself.

CHAPTER 5

Artificial memories and sensory implants: The hunt for the neural code

In 2016, a Silicon Valley start-up called Kernel emerged from "stealth mode" to publicly announce that they were building a prosthetic memory—a brain-implantable microchip that would not only help people with traumatic brain injuries regain their ability to recall information but also eventually help the rest of us become more intelligent. The possibilities were limitless, if you believed Kernel's founder, Bryan Johnson, who had just bet $100 million on the idea. "Could we learn a thousand times faster?" Johnson said at the time.[1] "Could we choose which memories to keep and which to get rid of? Could we have a connection with our computers? If we can mimic the natural function of the brain, and we can truly work with the neural code, then I posit the question—what can't we do?"

If you had been reading the science journals and tech press, you might have thought Kernel's plan was airtight. The rate of progress in brain implants over the previous decade had been astonishing, and Johnson had tapped into a growing body of apparently promising academic work. He had plucked one of the world's most prolific biomedical engineers, Theodore Berger, from the University of Southern California to lead the project as its chief science adviser. Berger had been working for twenty years on writing electrical signals into the neurons of rats and primates. He had just created an algorithm that could decipher the code sent by one part of the brain to another, and in so doing had apparently improved several rats' ability to form short-term memories.[2] Now, with an infusion of Kernel's cash, it was time for human trials. The Matrix was upon us.

Or was it? The belief that the right kind of implants can overwrite our normal brain activity has become practically an article of faith among the technocracy. "The future of the human race depends upon our ability to learn how to read and write our neural code," wrote Johnson in a subsequent *Medium* post.[3] But why? And where did it come from, this idea that we'll soon let academics and tech companies program our minds like a PC? The strange and twisting tale of Kernel's memory chip turns out to be an excellent parable, illustrating the profound limitations of our modern understanding of how the brain works. To understand why, we need to take a little dive into the idea of the "neutral code."

From heartbeat to neural code

The muscles in the heart either respond to a stimulus or they don't; that much had been clear to scientists since the 1870s. The rate of a heartbeat can vary but the beats themselves do not: there are no small or big heartbeats or half heartbeats. A heart either beats or it doesn't. Similarly, in early experiments, if you stimulated a muscle fiber, it either twitched or it didn't; there seemed to be no such thing as a half twitch. This was why du Bois-Reymond called it "all or nothing." For the heart, this binary made sense, because a functioning heart has one job: it just needs to beat.

But how could nerves and muscles use this same system to transmit more complicated information to and from the brain? How could they vary the content of the information they were carrying if all they could do was either fire or not fire? Nerves and muscles were clearly capable of acting on much more complex gradients of information. For example, you can choose to flex your arm either lightly, incompletely, or max out to exhaustion. And we are all familiar with the feeling of an initial sensation like sitting down in a chair or putting on a soft jumper, which after a while attenuates so we stop feeling it at all. Actions and sensations like these are hardly "all or nothing."

To find out whether muscles and nerves were really part of this all-or-nothing club, at the start of the 1910s the Cambridge University engineer and electrophysiologist Keith Lucas dispatched the usual complement of frogs. He was able to confirm that muscle fibers responded only when a stimulus was strong enough to exceed a certain threshold.

So all muscles obeyed the same binary rule—they either twitched or they didn't. Was the same rule true for nerves? And if it was, how on earth were they able to deal with complicated information?

Two problems blocked further understanding. The first: wherever you find them, nerves and muscles are not single wires, but bunched together in cables. These are a bit like the strands of fine wire ferrying signals along the ocean floor between continents. The signals aren't sent in one wire but separately along individual fiber-optic lines cabled into tight bunches of varying thickness. Similarly, the nerve "cables" in the body vary in thickness, some being very thick (like the spinal cord), others consisting of maybe a few tens of nerves.[4] The brain sends the message to the muscle, via the nerve, that it should contract. So whenever you try to listen to them, you'll get the din of loads of different neurons shouting over each other. Isolating an individual neuron to listen to its monologue was out of the question: first because it was surgically impossible to unpick a single (live) nerve from its fiber, and second because an instrument to detect its action potential didn't exist.

Even when listening to the loudest multi-neuron fibers in concert, you weren't listening to their natural conversation. All the way back to Galvani, every nerve or muscle signal ever measured had been "induced" by artificially stimulating the nerve to fire by applying an electrical zap. (I guess if we're going to stay with this metaphor, this is the equivalent to giving the nerve a giant static shock and listening to its enraged scream.) This method placed limits on how much you could learn about how the nervous system really worked in the wild.

Like any good physicist, the first thing Lucas did was find someone clever to take over the grunt work in his lab at Trinity College: a young PhD student in physiology, Edgar Adrian. Adrian's assignment: find out how the nervous signal is conducted and whether it also obeyed the same all-or-nothing principle Lucas had found in muscles.

They started by narrowing down the number of nerves in a muscle fiber that they had to contend with. Lucas found a muscle in the frog that was innervated by only ten nerve axons. When he stimulated it with an electrical zap, he found that the resulting muscle contraction depended on the intensity of the applied jolt. But that wasn't true of the individual nerves. They responded the same no matter the intensity of the electrical jolt: either firing or not. More stimulation caused more of the nerve fibers to fire, and that was what changed the amount of muscle contraction. The binary message in individual nerves never changed.

Here was solid evidence that nerves obey the same all-or-nothing law as muscle.[5] But the team's quest was disrupted by the First World War. Lucas left the lab and joined the Royal

Aircraft Factory, putting his engineering skills to use for the war effort by devising new compasses and bomb sights. While testing one of these devices in 1916, he died in a mid-air collision. When Adrian returned to Cambridge after his mentor's death, his obsession with Lucas's question had deepened. *How* was he going to listen to an individual nerve impulse? No one had made a machine powerful enough to record the signals themselves, but could you make something that amplified them sufficiently for existing machines to record?

During the war, Adrian's American friend Alexander Forbes worked on wireless radio receivers, early radar tools, and new devices called vacuum tubes which could enhance audio signals. The war had made them cheap and easy to get your hands on. After the war ended, Forbes used them to make a new amplifier, stuck the thing onto an Einthoven string galvanometer—and voila. A nerve's infinitesimal action potential could now be amplified by an unprecedented factor of fifty, which rose over the next few years to 7,000.[6] Such a great device—now if only he could find a way to listen to a nerve bundle in its natural state, not one that was being artificially jolted into firing by the application of electrical stimulation. Adrian got himself the blueprints to build a device of his own, and put in an order for frogs.[7]

The trick would be finding a scenario where nerve firing was predictable enough to be caught in situ and recorded. One day, he was recording the "resting" state in a frog muscle. This was intended to provide a silent baseline against which to compare the natural signals he hoped later to find. The frog leg was just hanging there, not doing anything and not being stimulated. Obviously no signal should have been present. And yet, every time he tried to get a good recording of this baseline resting state, the same annoying, unaccounted-for noise interfered, the same kind of oscillations that he had got when actively stimulating the muscle. The interference started to drive him around the bend, and Adrian laid the frog down on a glass plate—instantly, the mysterious signal stopped. He picked up the frog, letting its legs dangle once more. Signal. He put it down. No signal.

That was when Adrian suddenly realized what he was seeing. He understood the nature of the signal he had been detecting: the nerves attached to the legs were alerting the central nervous system that they were being stretched. He had found the signal they used to transmit this complicated information.

Now he'd need to find a way to record a single one of these signals as it traveled along a single nerve. Adrian set to work, and in 1925 he and his colleague Yngve Zotterman managed to reduce a muscle group to only one single strip, within which remained only one single muscle, within which remained only one single nerve. This sensory neuron was tasked with communicating just one thing: how much the muscle was experiencing the sensation of stretch. "Under strong emotional stress we hurried on," Zotterman wrote, "recording the nerve's response to varying degrees of stimulation." The signal they recorded from the single nerve was a series of clean, steady blips—the sound of a single, undiluted action potential. The blip was always the same. It never got bigger or smaller no matter how it was stimulated. The only thing that ever changed was how frequently it fired. Pull the neuron's muscle taut, and the blips became frequent and numerous. Give the muscle slack, and they slowed. When the muscle was perfectly at rest on the glass plate, there were no blips at all. Zotterman and Adrian both realized that "what we now saw had never been observed before and that we were discovering a great secret of life, how the sensory nerves transmit information to the brain."[8] It was a massive lightbulb moment—they were the first to finally figure out how the brain gets information from the limbs. They had cracked the code of how these blips tell the brain useful things about the information in the environment. Stretch: many frequent blips! Stop stretching: no blips. Something about this encoding sytem seemed awfully familiar.

The war, with its years of effort that involved breaking code and intercepting transmissions, had given Adrian a new conceptual lens through which to understand what he was seeing.[9] The mechanism he had found in the nerve's information transmission looked like a kind of bioelectric Morse code.

Nerve impulses, and the nervous system in general, had been described in terms of information communication since the invention of the telegraph in the previous century. But when Adrian discovered that the nerve impulse was just a series of variable-in-time brief pulses (Morse code without the dashes), he was struck by the way this limited signal was able to pass on complex information (the sensation of stretch). "In any one fiber the waves are all of the same form and the message can only be varied by the changes in the frequency and duration of the discharge. In fact, the sensory messages are scarcely more complex than a succession of dots in the Morse code."[10] You can see a similar change in the language of Zotterman's accounts. Years later, recounting the frustration of previous experiments before they were able to isolate that single neuron, he wrote that "it was as if we were tapping a telegraph cable with many lines simultaneously in transmission. It did not permit any reading of the code."[11]

Adrian's scientific and popular writings introduced the concepts that began to define the common perception of the nervous system and, by extension, of the bioelectric signal and its function: messages, codes, information.

This idea of a *code* began to bleed from single neurons into the idea of how whole nerve systems used action potentials to translate the outside world for interpretation by the brain. Now that there was a code being used by the peripheral nervous system to send messages to the brain, the next thing Adrian wanted to understand was how the brain received those signals—how it translated the Morse code back into a language it could understand. Was the brain a "central station" that decoded the signals into experience, as Adrian intimated in the lecture he delivered when accepting the Nobel Prize for this work? In which case, "we could tell what someone was thinking if we could watch his brain at work."[12]

Even before delivering that speech, Adrian had started combing the literature for any explanation. And while he didn't find it, he did discover a possible *way* to find it: a new machine recently invented by the German neurology professor Hans Berger. Its findings were of "exceptional interest" to Adrian, and he and his colleagues were shocked that no one had ever tried to repeat them.[13]

Hans Berger's quest for the brainscript

Almost ten years before Augustus Waller stood his first bulldog in saltwater, the Manchester physiologist Richard Caton was looking at similar rhythmic readings he had obtained from putting electrodes on people's scalps. Unlike Waller, Caton knew the significance of what he had just found. In the years since the electrical nature of the action potential was established, speculation had been growing as to whether the processing machinery in the brain might have its own electrical signature as well. In 1875, Caton discovered a "feeble current" emanating even when there was no muscle activity—not in keeping with scientific consensus of the time, according to which muscle motion should have been the only thing capable of generating measurable brain activity. And yet here was Caton's patient sitting perfectly still, emanating like a beacon.

Nearly fifty years later, his work was exhumed by Berger, then the director of the psychiatric clinic at the University of Jena.[14] Outwardly, the man was strict and charmless in the job.[15] His heart was elsewhere; he had been toiling in secret since the 1890s on a project of immense personal significance. It all went back to an accident he suffered in a military training exercise as a young man. In 1892, pulling heavy artillery on horseback, Berger was thrown off, his head landing inches in front of the wheel of an approaching artillery gun. The cart had stopped at the very last second—by all rights Berger should have died. When he returned to the barracks that night, shaken by the experience, he found a telegram from his father asking if he was all right. The reason for the inquiry: right at the time of his accident, his older sister was overwhelmed with an inexplicable feeling of panic and pleaded with her father to make sure nothing had happened to Hans.

Berger couldn't square the experience with science. What could explain such an extraordinary coincidence? He could only conclude that the sheer intensity of his terror had assumed a physical form external to his mind, and had been transmitted somehow instantaneously to his sister. Berger was determined to find the psychophysiological basis for mental telepathy.

In 1902, he discovered Caton's work on detecting the brain's electrical currents with an electrometer. After twenty more years of trying to find commensurate signatures in the brain, he finally got his hands on a string galvanometer. His first experimental subject was a seventeen-year-old college student named Zedel, who had been left with a large hole in his skull after the removal of a brain tumor. Berger attached Zedel's electrodes to a string galvanometer he had borrowed from the university hospital, where it was normally used to do early versions of the ECG. Suddenly, there they were—electrical tracings like the ones Waller had got from the heart, clear as day, but this time from the brain. Evidence of electrical brain emanations at last.

But the patterns he picked up from the brain were much more varied, faint, and noisy than the signals the clinical device had been able to detect from the heart, and therefore much harder to analyze for coherent motifs. Berger ordered an even bigger galvanometer. For five obsessive years he painstakingly tweaked the apparatus to tease out meaningful patterns from all the other interference that obscured them: the slightest bodily movements, the heartbeat, and even the pulsations of the brain's own blood flow.

By 1929, he had developed his new equipment enough to produce hundreds of recordings from patients with skull defects, epilepsy, dementia, brain tumors, and other disorders, as well as from healthy controls—himself and his son.[16] All showed consistent patterns in the waveforms; they were the same across many kinds of people. More intriguingly, they changed in similar ways. Their shape changed when you were paying attention versus when you had your eyes closed, for example. They changed when someone with epilepsy was having a seizure. It appeared that the shape of these waves did indeed tell us something about the internal processes of the brain. Finally, he had amassed enough proof to convince himself that he had devised a "brain mirror" to reflect the mental activity of the brain. He called his new tool an electroencephalogram, and it was the inexact first EEG—a device that could let you eavesdrop on the electrical activity of the brain. Almost five years after recording, Berger finally worked up the courage to publish his results.

Perhaps he had been wise to be reticent. The reception was icy, his paper mostly ignored. Thanks to his secrecy and the dour, uninspired reputation he hid behind, no one could believe that this little man had found anything groundbreaking. Many of his German contemporaries openly doubted that the oscillating waves he claimed to have found even originated in the brain. At a conference in Paris, while Berger explained the projections of his EEG graphs in a darkened auditorium, half the audience simply walked out.

Adrian, however, saw potential in Berger's work. He immediately began to work on it in his own lab, replicating and expanding it.[17] Berger found, for example, that the resting activity of the brain forms a pattern of wavelets he named the alpha rhythm, which hums along regularly, reliably producing between eight to thirteen of these little sawtooths every second. Intense mental activity changed the rhythm—the faster, more irregular waves he called beta waves. Adrian publicized Berger's work widely, going so far as to try to rechristen the alpha wave "the Berger wave."[18] He even staged a demonstration for the Royal Society in which he took traces of himself thinking in public—changing the shapes of his own emitted oscillations in real time.[19] No bulldogs were involved.

Reading EEG patterns now began to allow American technicians to distinguish sleep from waking states, focus from inattention, and even healthy brains from those with neurological disease.

In Germany, the public imagination was beginning to enter its fever phase, and in the late 1920s and 1930s, the EEG's ability to make recordings of the electrical activity of the human brain began to kindle far-reaching speculations about the imminent deciphering of the brain—and consequently, the mind. A German journalist wrote enthusiastically that "today the brain writes in secret code, tomorrow scientists will be able to read neuropsychiatric conditions in it, and the day after tomorrow, we will write our first authentic letters in brainscript."[20]

This enthusiasm would not last. At some point, the optimistic tone disappeared, leaving only the worst-case scenarios. A radio program investigated the worrying "electrophysiological problems of the future."[21] Editorial cartoons captured the attitude of the average German at the time: one speculated that the addict of the future would be addled by electricity instead of cocaine and morphine; and another—a brutalist depiction of a person's exposed brain irradiated by waves that shone out through his confused eyes—envisioned brainwashing by a surveillance state. The cartoon's caption: "Intensification of suggestive forces by the supply of electrical oscillatory energies into the brain."[22]

Then there were the enterprising opportunists who seized on EEG, whose schtick you'll find quite familiar by now. Berger's discovery ushered in a booming market of quack medical appliances. One buyer asked Berger for advice on using EEG to assess the temperament of his new horse. The head of a women's clinic at Tubingen tried to use EEG to establish neural signatures of pregnancy.[23] Berger was furious about all of it.

Outside Germany, by 1938 the tool was being used all over the world. It was particularly useful for diagnosing the patterns characteristic of epileptic seizures, sleep stages, and drug responses. The study of EEG advanced at a particularly breathtaking speed in the US, where wartime tech and American open-mindedness was leading to breakthroughs in theory, devices, and praxis.[24] When new EEG labs broke ground at universities, the ceremonies attracted luminaries from all over the country. But this was not an era known for the open sharing of scientific knowledge between Germany and other nations, so Berger had no idea how much his EEG was already changing the face of neuroscience in the US. He only saw what he fulminatingly called the "ballyhoo" his creation had brought about in his own country. In 1941, just as Adrian was writing up his letter recommending Berger to the Nobel committee, the latter, mired in despair and depression, took his own life.

After seventeen years of progress in EEG technology, the field stalled for another four decades. During this time, we decided that we'd rather send electricity into the brain than decipher the codes hiding in the natural kind.

How we decided the brain was a computer

At the dawn of the computer age—as engineers began to assemble the very first room-sized computing machines—those computers were also being built (and conceived of) as a kind of brain. In 1944, the electronics manufacturer Western Electric, in a glossy *Life* magazine ad for its new anti-aircraft guidance system, declared that "this electrical brain—the Computer—thinks of everything." The next logical leap was inevitable: if a computer is a kind of brain . . . might the brain be a kind of computer?

The American neurophysiologist Warren McCulloch entertained that possibility. He was already familiar with Adrian's search for the messages hidden in the nerve's all-or-nothing firing rates. And now, as he became familiar with the binary coding that was the basis of computing, he identified a possible correlation. In computers, the binary choice is between "a statement that is either true or false": 0 or 1. In the brain, "the neuron either fires or it does not." Could all-or-nothing neural firing be the brain's version of binary code?

The vocabulary in the two disciplines soon overlapped. During the following years and decades, McCulloch and his colleagues in many disparate disciplines seeded their descriptions of how the nervous system worked with electrical engineering terms. Neurology adopted terms like "brain circuits." Electrophysiology adopted terms like "circuit," "feedback," "input," and "output" into descriptions of how the nervous system worked. Increasingly, the line blurred between the code you write into a computer to program it, and the idea that brains were subject to similar governance.

All this intermingling soon bred a formal new school of thought: cybernetics, an idea that emerged from the Second World War, was seen as a science of communications and automatic control systems, relevant for both machines and living things. But for its most ardent disciples, it was also a means toward mind control. The main idea in cybernetics is that if anything a human (or any animal) perceives and experiences is just code routed through the brain by the nervous system's circuits, you should be able to control a human mind as surely as you can control a machine. It wasn't just scientists who succumbed to the cybernetics craze—this new understanding soon fully penetrated the zeitgeist. Engineers built robots whose operating systems purported to model the human brain, and imbued them with consciousness-like qualities thanks to their ability to "perceive light" or return to their charging stations of their own accord.[25] By the time Norbert Wiener published the hugely influential book *Cybernetics: Or Control and Communications in the Animal and the Machine* in 1948, the idea was already wildly popular and the book became an international bestseller, despite containing, as the historian of science Matthew Cobb has pointed out, "vast tracts of equations that were incomprehensible to most readers (and were full of errors)."[26] In other words, the idea was so compelling, there wasn't much point in bothering with whether it was based in fact. The idea that we should be able to drive an animal like a robot by simply activating specific circuits of neurons was too good to verify.

But with what tools would it be possible to control the human circuitry? Scientists went back to a tried-and-true method—shocking people. (Even Edgar Adrian had had a brief dalliance.[27]

During the First World War, while finishing his medical studies in London, he and his colleague adapted "torpillage," a kind of electro-therapy popular in France and Germany, to cure British soldiers of shell shock and get them back to the front as fast as possible.[28] When he realized that soldiers relapsed more often than they improved, Adrian abandoned the practice in 1917 and returned to the work that would lead to his Nobel.)

At first, electro-therapists administered jolts of electricity to the whole brain, with no luck. But what if, instead of blasting a person indiscriminately with electricity, you targeted the shock at a specific brain circuit? Specific brain areas were a hot topic. In the 1940s, while looking for parts of the brain responsible for epileptic seizures, the neurosurgeon Wilder Penfield found a remarkable clue that there were areas deep in the brain responsible for very specific experiences and memories. Before cutting out the bits of brain tissue that generated the epileptic symptoms, Penfield would first locate the problem area by electrically stimulating several parts of the deep brain. Strange behavior would ensue. His patients might suddenly start singing lyrics to a song they hadn't heard since they were children; they might say they smelled some powerful phantom scent. Electrical activation of some brain areas clearly brought sensations out of the dark closet of the deep brain and into the light of day.[29]

Emboldened by these hints about what was encoded in brain circuits, other scientists opened up people and animals and stuck electrodes into them for more precise control. Early approaches focused on the brain's pleasure centers and reward circuits. This approach had powerful consequences. An electrode that hit the right spot in a rat's brain would cause the animal to do absolutely anything to stimulate itself, including staying awake doing nothing else for twenty-six hours.[30]

The discovery of such a control switch in the mammalian brain led to exactly the kind of ethical trainwrecks you might expect. In the late 1960s, a patient came to the office of Robert Heath, a psychiatrist at Tulane University in New Orleans. This patient was desperate to be cured of his homosexuality, understandable given the cultural attitudes in 1960s Louisiana. By the time the patient—Heath referred to him as B-19 in his records—sought professional help, he was suicidal. So Heath implanted his patient with a stimulator with the view to reorienting his desires toward women. While B-19 controlled the self-stimulator, Heath instructed him to watch unlimited heterosexual pornography in the lab.[31] Heath reported that "B-19 stimulated himself to . . . almost overwhelming euphoria and elation, and had to be disconnected, despite his vigorous protests." After some time, B-19 wanted to try it in the flesh, and Heath procured a prostitute to visit the lab. The psychiatrist's clinical observation: "The young lady was cooperative, and it was a very successful experience."[32] The long-term effects, however, were less conclusive. While B-19 did go on to have a long-term heterosexual relationship, he never stopped having sex with men. Simply zapping the human reward circuitry, it seemed, had its limits. So did the public's patience for Heath's work in this area, which in 1972 was denounced as a "Nazi experiment" by a local magazine, a judgment that sent his career into a terminal plunge.[33] But his work had already been overshadowed by something much more exciting and media-friendly: instead of a "go" button, something that *stopped*.

José Delgado was a Spanish neurophysiologist at Yale University who, in his formative academic years, explored the neural roots of aggression, pain, and social behavior, just as cybernetics was gaining steam. It was under this framework that he began his research on electrical stimulation in animals. He was soon expertly building custom microelectronics to implant into the brains of cats, rhesus monkeys, gibbons, chimpanzees, and bulls.[34]

In the mid-1960s, Delgado went to a ranch in Córdoba, Spain, to investigate the areas of the brain where neural activity was related to aggression. For his experiment, Delgado chose a fighting bull called Cayetano and another called Lucero. Each weighed well north of 500 pounds.

Delgado inserted a battery-powered electrode into an all-purpose area in Lucero's brain that was involved in everything from movement to emotion. Then he got him angry. When the bull charged, at the last moment Delgado pressed a button on a radio device that remotely turned on the stimulating electrode, electrifying Lucero's caudate nucleus and causing the bull to stop short.

The grainy photograph depicting the famous experiment has probably made its rounds through every undergraduate neuroscience seminar in the world. Delgado cuts an improbable figure in slacks and a professorial V-neck pullover over a collared shirt, facing down a beast that is charging toward him in a fenced enclosure. He is holding up something that looks like a portable radio with an antenna—standing apparently unfazed in front of a bull that seems to have skidded to a halt so short, its rigid hooves are barely visible behind a cloud of dust.[35]

Lucero's implant did not only stop the bull from charging. If the bull was eating when Delgado pushed his remote control, he would stop eating. If the bull was walking, pushing the button would cause him to stop walking. It seemed that Delgado had found in this brain area something like a universal "stop" button. The sudden transition from rage to peace led *The New York Times* to call the experiment a "deliberate modification of animal behavior through external control of the brain."[36]

Delgado continued to explore the control of aggression, passivity, and social behavior using implants in humans, chimps, cats, and many other animals. In 1969, he published a book discussing his experiments and their implications, called *Physical Control of the Mind: Toward a Psychocivilized Society*. It achieved instant notoriety, if only because of its final chapter, in which Delgado—whose cybernetic ethos had been shaped by five months in a concentration camp—declared that humanity was on the verge of "conquering the mind," and should shift its mission from the ancient dictum "Know Thyself" to "Construct Thyself." Used wisely, he insisted, neurotechnology could help create "a less cruel, happier, and better man."[37]

No one could propose implanting a stop switch into a human brain for such speculative reasons. But soon a much more compelling use case presented itself.

A pacemaker for the brain

It was an otherwise quiet morning in 1982, and a patient named George had just been admitted into a psychiatry unit with a diagnosis of catatonic schizophrenia. The designation had been applied not because it fit, but because nothing else did. The patient was unresponsive, but he still seemed alert, a combination outside all the existing frameworks available at the time. The psychiatrists were sure the patient had a neurological disorder, and the neurologists were sure the patient had a psychiatric disorder. Eventually, the chief resident ran to the office of the director of neurology, Joseph Langston.

Langston set aside his coffee, looked up from that morning's EEG reports, and began conducting his own tests and consultations. Initially, he concluded George exhibited all of the symptoms of advanced Parkinson's disease, a cruel neurodegenerative disorder whose iconic symptom is a trembling so violent, a person can't so much as hold a cup of water, and which progresses after many years into a rigid freeze. But Langston knew the diagnosis couldn't be right for two reasons. The patient was only in his early forties, which was about twenty years too young for a diagnosis of Parkinson's. And instead of manifesting gradually over the course of years or even decades, his (apparently) end-stage symptoms had appeared literally overnight.

The mystery only deepened further when they found George's girlfriend frozen in the same state despite being even younger—she was only thirty. Eventually, the team located five more identical cases. It took some sleuthing and luck, but Langston and the police eventually uncovered the common factor: each of the patients had recently used heroin—or at least what they thought was heroin. When Langston's group got their hands on some samples, what they found wasn't heroin at all: the street chemists had mistakenly synthesised a compound called MPTP. A search through the medical literature revealed several existing studies on MPTP, and their findings didn't bode well for the couple. By destroying an area deep inside the brain called the substantia nigra, MPTP had been found to create irreversible symptoms that mirrored Parkinson's—notably that rigid freeze.

The discovery of that relevant brain area was consequential. During the 1970s, a few neurosurgeons had been experimenting with implanted electrodes for chronic pain and epilepsy. They drilled open the skull and pushed a penetrating electrode deep into the grey matter. It was a promising solution to the big problem that had driven psychosurgery out of business: unlike the traditional approach of burning or cutting out bothersome bits of the brain, "electrical lesions" were adjustable as well as reversible. If you were applying too little electricity, you could dial up more; if you were putting in too much, you could dial it back down.

As they did this, doctors began to notice two patterns among the patients exhibiting troublesome symptoms: first, the electrical stimulation alone was sometimes enough to blunt the symptoms. Second, the faster the electrical stimulation pulsed, the more the patients improved.

These patterns were intriguing, but the electrodes couldn't be implanted for take-home care. Like Hyman's earliest pacemaker, they were hooked up to an unwieldy external apparatus, a large power source which was connected to the electrodes that stuck out of the head by wires.[38] Moreover, no one had done any large clinical trials to verify that stimulating some particular brain area worked for everyone, or if it was just bespoke to the individual patient. The only assessment of whether it worked at all was the assurance of the implanting surgeon.[39] However, all the growing interest had led Medtronic to work on adapting their pacemaker to make it suitable for brain implantation. They sent their experimental devices to specialist centers, and even trademarked the term DBS ("deep brain stimulation"). Their devices were still limited to small, one-off experiments. But that all changed when George's revelatory case study made it to the desk of Alim-Louis Benabid at the University Hospital in Grenoble, and he connected the dots.[40]

Benabid was among the few psychosurgeons still using implanted electrodes to identify the correct brain area in advance of psychosurgery. He had become fascinated with the clear and obvious effects he saw in his Parkinson's patients: symptoms would quiet in real time in the operating theater. When he grasped the significance of George's case study, he obtained Medtronic's new brain pacemakers and implanted a handful of patients. The improvements were dramatic. Stopping the faulty neurological code emerging from that part of the brain settled the trembling, allowing patients to once again move their limbs as they wished. Medtronic hired Benabid to design massive trials. Now, instead of unscientific zaps accompanied by an unconvincing thumbs-up from a clinician, here was a disease with extremely well-spelled-out symptoms, a relevant brain area, and a robust response to an already approved medical device.

Medtronic had been desperate to find a way to expand its hugely successful pacemaker business. In Benabid's work, they saw a new opportunity. And in trial after trial they found the same dramatic effects: start the current flowing through those deep electrodes, and trembling subsided instantly. People who before the surgery could not hold a cup of tea were now able to confidently brew themselves a whole pot. EU regulators approved the implant for Parkinson's in 1998, and the Food and Drug Administration (FDA) approved the treatment in the US in 2002. One of the doctors who began implanting his patients hailed it as "providing a new life." The pacemaker had moved into the brain and deep brain stimulation was born.

More than 160,000 of these "brain pacemakers" have been implanted to mitigate the disabling muscle spasms of people with Parkinson's, essential tremor, and dystonia.[41] As brain surgeries go, it is actually pretty straightforward. First, drill two holes in the skull. Next, push two metal electrodes, each about the dimension of dry spaghetti, into the region of the brain responsible for the symptoms. Finally, snake the wire through the head and down the neck until it reaches, implanted under the skin near the collarbone, an object about the size of a stopwatch. That's the pacemaker! But now it's sending a current whose pulse and amplitude, over the following couple of weeks, a technician will adjust until the symptoms subside.

In big clinical trials with many participants, knowing which brain area to overwhelm with electricity has allowed surgeons to successfully short-circuit faulty signals in these broken regions. What other brain areas were amenable to freezing in this way, to widen the net of conditions controllable by a pacemaker? Many small trials offered clues to the next big ailment.

In 1999, researchers at the Catholic University of Leuven (KUL) in Belgium implanted DBS electrodes into an area called the internal capsule in four people with severe obsessive-compulsive disorder. Symptoms improved for three.[42] More trials on other ailments followed in rapid succession—again small, investigational studies, often only with ten or fewer participants. But despite their small size, these trials generated dramatic headlines, as my colleague Andy Ridgway noted in 2015 in *New Scientist*.[43] DBS allowed a thirteen-year-old autistic teenager to speak for the first time.[44] It freed people with Tourette's from bone-breaking physical tics. It allegedly stopped obese people from overeating and anorexic people from undereating.[45] The small trials proliferated—what else would yield to a brain pacemaker? Anxiety? Tinnitus? Addiction? Pedophilia?[46]

Medtronic bet on depression. They weren't alone in thinking this was promising. The idea had been germinating since 2001, when neuroscientist Helen Mayberg had the idea to investigate DBS for intractable depression (the kind that refuses to budge no matter the treatment). DBS, she told me when I met her at the International Neuroethics Society symposium in San Diego in 2018,[47] "seemed to block abnormal brain function in Parkinson's, so we wanted to block our own depression-specific area." Mayberg focused on a brain region known as Brodmann's area 25, which has been dubbed the brain's "sadness center." Too much activity here, Mayberg and her colleagues thought, was driving symptoms like negative mood and that characteristic lack of a will to live. What would happen if you froze those neurons? Four of her first six patients saw dramatic improvement.[48] Twenty more small trials recorded improvement rates as high as 60 or 70 percent. "People come out of that very dangerous state, and stay well," Mayberg told Ridgway. "They just get back on the bus." In other trials all over the world, depression lifted similarly.

After enough such intriguing results had piled up, St. Jude

Medical—a Medtronic rival—took the plunge and bankrolled a major trial. It seemed all but guaranteed to culminate in the first new commercial application of DBS since Parkinson's. Two hundred participants at more than a dozen medical centers got the implant. The buzz was immense. And then, after six months, the trial stopped. Gossip spread among industry insiders that it had failed an FDA futility analysis, meant to make sure spendy trials get kneecapped if they are clearly wasting time and money. There were stories of terrible side effects and a suicide attempt.[49] The implication—bad for the future of the technology—was that there was no difference between a placebo and an implant.[50]

When all the drama and recriminations settled, the story ended up being a lot weirder and less straightforward than the first reports suggested: as the journalist David Dobbs concluded in a deeply reported postmortem analysis published in *The Atlantic* in 2018, it appeared to be a case of a treatment that actually seemed to work being sabotaged by a trial that didn't. For the people in whom it worked, it was a gift, yielding immediate results that were so dramatic as to be nearly magical. "What did you do?" would be the response of the awake patient, still in the operating room, the moment the stimulator was turned on. And when that happened, the results were enduring. "If you got better, you stayed better, given continued stimulation," Mayberg told several media outlets in the wake of the trial. The same pattern held true for obsessive-compulsive disorder: those patients who had responded well in a brain stimulation trial at the Catholic University of Leuven still had their obsessive-compulsive disorder under control after fifteen years. "It was like someone spring-cleaned out my brain and took out all the unnecessary thoughts," another participant told Alix Spiegel, host of the National Public Radio show *Invisibilia*.[51]

But it was just not possible to predict who would have a miracle and who wouldn't. There were also some strange side effects.[52] The deep, ancient areas of the brain targeted for electrification in depression and Parkinson's are involved in much more than motor and mood control. They are implicated in learning, emotion, and reward—which is to say, addiction. Interfering with these had unpredictable consequences. This was the case for one Dutch man being treated for severe obsessive-compulsive disorder, anonymized by his doctors at the University of Amsterdam as Mr. B. His new brain implant had been working for just a few weeks when he chanced upon a recording of Johnny Cash's "Ring of Fire." In the five decades before the twin electrodes penetrated his deep brain, he had never been especially moved by music—he was the kind of guy who claimed to like both the Beatles and the Rolling Stones. The day Johnny Cash's voice hit his newly electrified pleasure centers, however, all that changed. From then on, no other music was allowed. Mr. B bought every Johnny Cash CD and DVD he could get his hands on. But when the electrical stimulator was off, he could not for the life of him recall what was so important about Johnny Cash.[53]

Not all side effects were as endearing, though. People with Parkinson's implants have reported an increase in impulse control disorders, such as excessive gambling and hypersexuality.[54]

This reflects a somewhat uncomfortable open secret about DBS: despite all the complicated talk about the function of precise areas of the brain, no one is sure exactly how DBS works.[55] As recently as 2018, academic reports described DBS as an effective but "poorly understood" treatment, even in Parkinson's and the other motor diseases it has been approved to treat for decades.[56] "If you think of neurons executing the neural code as playing a melody on a piano," says Kip Ludwig, a former director at the US National Institutes of Health (NIH), "then DBS is like playing that piano with a mallet."

There's a limit to this approach. Electrifying specific brain areas could broadly control some maladies, but it was impossible to get truly granular enough to reliably hit a target as ephemeral as depression. We needed to know exactly what was happening in the brain's code as a response to those zaps.

For that, we needed to decipher the neural code.

Reading the neural code

By the 1970s, Francis Crick had grown bored of molecular biology, even though he'd essentially invented it. He was looking to solve his next great mystery. If cracking the blueprint for life had been exciting, how about cracking the secret of consciousness? So in 1977, he left Cambridge for the Salk Institute in California, where he turned his attention to what he considered a deeply unpromising approach to neuroscience. He demanded new "theories dealing directly with the processing of information" and ways to tie behaviors and actions to the neural firings that accompanied them.

In 1994, he summarized his research in *The Astonishing Hypothesis*, a slim volume that would have an explosive impact on neuroscience and philosophy. "One may conclude," he wrote, "that to understand the various forms of consciousness we need to know their neural correlates."[57] He further argued that all those things we think or feel or see "are, in fact, no more than the behavior of a vast assembly of nerve cells and their associated molecules."[58] (He did not address how this was materially different from our identity being no more than the behaviour of a vast assembly of genes.) The subtitle—*The Scientific Search for the Soul*—made clear the book's ambition.

In the whole two decades before Crick's book was published, fewer than ten peer-reviewed scientific papers referred to the term "neural code." After the release of *The Astonishing Hypothesis*, however, neuroscientists increasingly turned their attention to trying to find the neural signatures of a vast array of behaviors and thoughts. For students of the sensorium, neural code was the new black.

Not that they knew what the term meant. Even as Crick was writing his book, the definition of the term was the subject of a contentious spat in neuroscience. Adrian's ideas that information could be encoded in the Morse code dots passed by individual neurons still had its adherents, but now there was a fresher idea. Brain plasticity—summarized by the axiom "neurons that fire together, wire together"—became ubiquitous thanks to its succinct explanation of how different neurons learn to work together as you learn different skills, from language to ballet. A real neural code couldn't possibly focus on the firing of single neurons, representatives of the new guard wrote in 1997, but had to take into account the way vast collections of different neurons fired together in synchrony to form a coherent pattern over time and space.[59]

It would be one heck of a hard thing to measure. By then, we had begun to grasp the sheer size of the brain—86 billion neurons. No tool was, or is (or possibly will ever be), capable of reading the activity of all those neurons at the same time. But as the twenty-first century approached, we had options.

The trusty EEG was still in use, a workhorse that had given us the different waves that could reveal focus and inattention, but also much more. Scientists spent decades using these readings to advance our understanding of sleep. Because EEG doesn't require opening the skull, just some electrodes on the scalp, scientists were able to obtain a lot of data from a lot of people.

The EEG evolved from its humble origins in Hans Berger's lab into skullcaps encrusted with dozens of electrodes that could read subtle variations in the chorus of the brain's billions of denizens. That helped drill down into the brainwaves, and the subsequent discovery of delta and gamma waves (in addition to Adrian's alpha and beta) helped researchers identify the different stages of sleep, yielding the familiar, ever-deeper stages of I–IV and dreaming sleep. Other work tied characteristic disruptions in these waveforms to sleep disorders and neurological disorders—even helped identify the location of brain tumors. Thanks to increasingly powerful computer processing power and better signal-processing algorithms, EEG could more finely analyze the brain's patterns. Depression was correlated with an overabundance of alpha waves in the EEG. In Parkinson's disease, there was a paucity of beta waves. Alzheimer's patients have been found to have a deficit of high-amplitude gamma waves. Various research papers reported a rainbow of emotions correlated to waveforms Berger couldn't have dreamed of.[60]

Another tool, electrocorticography (ECoG), could get deeper into the brain, but it would be amenable to a smaller population. It looked like a mat of electrodes placed directly onto the exposed folds of the brain, a bit like a doily on a side table, to record electrical activity from the cortex. It does require opening the skull, which is why it is rare to obtain these sorts of traces. The only human volunteers for this kind of brain reading have already had their skull opened for unrelated investigational purposes. Sometimes these individuals gave researchers permission to put the mesh on their brains and read the neural correlates of certain specific thoughts, but it still didn't let the researchers get next to any specific neurons.

For that, you needed an invasive brain-penetrating electrode. The first of these was approved to stick into human brains in the 1990s. It was called the Utah array, and it looked like a small metal square with ninety-six electrodes sticking out of it, a bit like a bed of nails for a ladybird. Nestled into brain folds, it could record the squabbling of many neurons talking to each other, or hone in on one in particular. But this reading was the most invasive of all: it required not only opening the skull (or drilling a hole in it), but also jamming the penetrating electrode through the blood–brain barrier, and running a lead out of the skull to power and listen to the array. The only subjects on whom it was considered ethical to test this device were animals—and later, people who had suffered the kind of irreversible physiological trauma that made this the last faint hope for the possibility of a research breakthrough that would help them.

By 2004, the theoretical neuroscientist Christoph Koch—a buddy of Crick's who profoundly influenced his ideas about the neural correlates of consciousness—predicted that thanks to these and other new tools, it would soon be possible to decipher the workings of the neural code to understand consciousness, language, and intentionality.

Around the turn of the twenty-first century, the evidence for this optimistic reading of the future was all around us in media reports. In 1993, an invasive electrode placed into the cortex of a woman who had been paralyzed by a stroke allowed a computer to determine where exactly on a square of letters she was focusing her attention. ECoGs detected the electrical signatures of people thinking in whole words: "yes," "no," "hot," "cold," "thirsty," "hungry," "hello," and "goodbye."[61]

So it seemed that, in line with Koch's predictions, you could indeed use the brain's electrical signals to peer into people's minds. By 2022, at least fifty peer-reviewed papers every year were invoking the term "neuron code." Many of these papers investigated what actions, thoughts, and feelings could be traced back to bio-electric signals in the brain.

That raised a new question: if you could read the brain's state by examining its electrical signals, what would happen if you changed them? Could you reprogram the brain?

Writing the neural code

On 22 June 2004, a little metal pincushion was pressed into Matt Nagle's motor cortex, specifically, into the region that controlled his dominant left hand and arm. After he had been paralyzed from the neck down in an accident, the neuroscientist John Donoghue had signed him up for a clinical trial called BrainGate, and implanted him with a Utah array. Eventually, Nagle was able to move a computer cursor with just his intentions. If he wanted to move the cursor to the left, the motor neurons in his brain fired how they normally would have to control his fingers. The Utah array picked up that signal, translated it into machine language to control the cursor, and the cursor moved left. That's how, in 2005, Nagle beat a *Wired* magazine reporter at the computer game Pong.[62]

Donoghue had much bigger plans. If these signals could drive a robot arm, why couldn't researchers find a way to take control of a real arm—namely Nagle's own? In 2005, he told *Wired* that his eventual plan was to "hook BrainGate up to stimulators that can activate muscle tissue, bypassing a damaged nervous system entirely."[63] It was ambitious and very exciting (if a bit Frankenstein): instead of trying to heal the spinal cord injury that had disconnected the limbs from the brain, the BrainGate implant would beam the electrical signals that drove intent to their intended endpoint directly, and so reanimate the limbs.

The idea was called a neural bypass, and within a decade, it was being demonstrated in a TED talk.[64] "The idea is to take signals from a certain part of the brain and reroute them around the injury—whether that injury is to the brain or the spinal cord—and then reinsert those signals back into the muscles to allow them to regain movement," Chad Bouton told the audience, pacing the stage like a TV-handsome talk show host. Bouton was the engineer who had developed the signal processing algorithms for the original BrainGate project. "But we still had not given [our subjects] movement," he said, frustrating his vision of helping people with life-altering spinal injuries to walk again. When, in 2008, Cyberkinetics, the company that owned BrainGate, failed, Bouton moved to the Feinstein Institute in Manhasset, New York, to begin work on his neural bypass. Under the auspices of a project funded by the US Defense Department, he joined a research supergroup with the science institute Battelle and Ohio State University, and in 2014 they implanted a computer chip into the motor cortex of a young man named Ian Burkhardt, who had become quadriplegic after a diving accident.

For Bouton and the others working on restoring fine motor control, deciphering the neural code was not a matter of counting the action potentials generated by every nerve the way Edgar Adrian did it. In a brain full of 86 billion neurons, you can't possibly pick out and analyze the spiking behavior of the billion or so involved in every motion. Bouton thought the thing to focus on instead was how groups of neurons synchronize the timings of their firing when any particular intention was recorded. He called it a "spatio-temporal" relationship. After harvesting this pattern, they would re-encode it into machine language, and this would animate a cuff of electrodes around Burkhardt's wrists. Instead of actuating motors on a robot limb (the way BrainGate had done), each electrode would stimulate tiny swatches of muscle in his own arm.

It wasn't exactly the way a brain signal would innervate the muscle, but the tortured mathematical transformations worked. With the help of the device, Burkhardt picked up a mug of water and lifted it to his own lips to take a sip. Ian Burkhardt had become the first person to use a chip to "reanimate" his own living muscles with the neural code harvested from his own brain.[65] The signals were precise enough to let him play Guitar Hero.[66]

But Bouton still wasn't totally satisfied. What good is it to be able to move if you have no sense of what you're touching? This was a practical question. "It seems so simple to you and me, but if you have no sense of touch in your hands, no pressure or slip awareness, you don't know if your grip is tight enough," Bouton told me when I visited his lab at Feinstein a couple of years later. Without that grip awareness you'd be just as likely to drop the cup as suddenly crush it and spill hot coffee all over yourself (and your lack of pain awareness would make you unaware that you've just given yourself second-degree burns that need a doctor). To avoid this, an implantee must train their full attention exclusively on gripping that cup from the moment they pick it up to the moment they let it go. "There was one guy with the implant," Bouton said, "he could pick stuff up, but as soon as he wanted to do or think about literally anything else, he'd drop whatever he was holding." Imagine if you had to do that every time you wanted a sip of coffee. And now imagine it's not a cup of coffee you're holding, but your kid's hand. All these activities of daily life would be meaningless without sensation. A half measure, thought Bouton.

The sensorimotor cortex, where sensation lives, is right near the motor area where intention lives. That's the good news. The bad news was that writing the correct pattern of electrical spikes into the brain to replicate the experience of sensation would be a much harder problem than reading existing signals as they fired.

Almost exactly six months later, a paralyzed volunteer called Nathan Copeland—working with a different research group at the University of Pittsburgh—lay blindfolded next to a five-fingered robotic arm. Each time a researcher prodded one of the robot's "fingers," Copeland identified where on his own hand he was feeling the touch. "Index finger," he noted as the researcher touched the machine's index finger. "Middle finger. Ring finger," and so on, and thus followed a long series of correct localizations.[67] In addition to the usual BrainGate-style motor cortex implants, Copeland had had two electrode arrays implanted into areas of his brain whose neurons respond to sensation in the fingers (each implant was about the size of a sesame seed). Every time the researcher jabbed the robot's finger, these little seeds sent electricity patterns into the correct neurons.[68]

This mechanism interested Theodore Berger (no relation to Hans), but instead of implanting sensations, his electrodes would spark artificial memories.

The memory maker

Berger's goal was to mimic the function of the hippocampus, a part of the brain where memories are processed and encoded. He had long been working on a chip that recorded whatever brain pattern corresponded to a behavior he liked, then fed it back into the brain using what he called a multiple-input/multiple-output (MIMO) algorithm to reproduce the behavior.

He tested MIMO by temporarily damaging a rat's brain, in a way designed to specifically block the hippocampus's ability to write memories, mimicking the effects of dementia. He had previously recorded the rat successfully doing a specific task. With the brain damage, the rat now couldn't repeat the performance. But when the damaged hippocampus was zapped with the MIMO patterns recorded earlier, the rat's performance on the memory task went back to normal—even though its brain was still damaged.[69]

Berger believed he had created a prosthetic memory (although many would dispute that characterization). He and his co-authors concluded that "with sufficient information about the neural coding of memories, a neural prosthesis can restore and even enhance cognitive processes." That wasn't all. Code taken from any one rat could be zapped into any other rat, which seemed to suggest that Berger had discovered aspects of a universal code that governs how all creatures form memories,[70] even rhesus monkeys.[71] This time, MIMO was tapped to intervene whenever it looked like a monkey was about to make a wrong choice. Monkeys that got the "right decision" code stimulation made better choices 15 percent of the time. This showed, Berger insisted, that MIMO was not specific to one particular animal, raising the possibility that one day when you're reaching for the fries, your good decision implant could give you a little "salad" zap.

Berger had long relied on grants from the Defense Advanced Research Projects Agency (DARPA), widely known as the US military's mad science wing, and his research dovetailed nicely with their efforts to understand the neuroscience of memory and traumatic brain injury (with a view to fixing those sustained from IEDs and other war injuries). The agency was funding a prosthetic memory device for implantation in a human brain,[72] but both the timeframe (too short) and the money (not enough) put human trials out of reach. Enter Bryan Johnson, who had recently pocketed $800 million from selling his online-payments company to PayPal, and was looking for something more exciting to invest in.[73]

When he discovered Berger's work, Johnson immediately dumped $100 million into the new start-up, Kernel, that would bring the memory chip to reality. There seemed to be no ceiling to what you could do by tapping the neural code. If all your senses ultimately boiled down to electrical signals hitting different areas of the brain, couldn't you make a memory from scratch by impersonating those signals?

The engineers said that all they needed was access to more neurons. Another group of researchers, who said they had transmitted a memory of a sensation, had used thirty-two electrodes. But Johnson told reporters that the plan was prosthetic memory implants containing nearly 2,000 electrodes, and that 5,000 or even 10,000 were achievable. The same year, SpaceX and Tesla entrepreneur Elon Musk proposed a brain implant that would read to and write from *thousands* of neurons simultaneously. (Not known for modest goals, Musk advocated using this implant to "co-evolve" with artificial intelligence.) It seemed like a fairly straightforward, linear progression: the more neurons you could manipulate, the more precisely you could write the neural code; the more precisely you could write the neural code, the more powerful the brain interface. Therefore, if you want to read and write more neurons, just add electrodes.

However, one does not simply "add" more electrodes (see Chapter 9 for more on this). Not long after Kernel snapped up Berger, Johnson convinced Adam Marblestone to leave his post at MIT's Synthetic Biology lab to become the company's chief strategy officer. But when he began to review Berger's work and Kernel's goals, Marblestone and his colleagues saw a potential problem. First, Berger was working with a device that had a total of sixteen electrodes, far less than even a Utah array. Second, saying the algorithm had restored memory might have been an overly generous interpretation of the experiment, as the tasks were both narrow and basic. "To say it meant you 'read the neural code' would be like saying you have cracked language when you only decoded the words 'yes' and 'hello,'" said Marblestone, "technically true but potentially overstated."

After a failed human experiment, the collaboration fell apart. Still, Marblestone didn't think this a reason to be overly cynical. "Given that we don't yet understand much about the neural code, and the lack of technology with which to densely read and write," he says, "we just don't know yet whether it's possible or not." But that's not a statement you can take to investors. After Johnson realized that there was no way to take what Berger had built and scale it up to build something anyone could or would buy, Kernel abandoned the memory chip.

In the end, it was the full appreciation of the hardware problem that drove Bryan Johnson to rethink his relationship with the neural code. "For a while, Kernel was looking at maybe building a medical device a bit like DBS," Marblestone says. "But we don't really know what can be done with DBS-type devices, beyond Parkinson's."

And that explains what happened next: Marblestone advised Johnson to abandon the idea of writing the neural code, and instead figure out the most interesting thing that could be done with the brain *without* opening the skull. Johnson listened, and Kernel settled on reading brain signals. They started building something that could measure other signatures of mental activity while the brain is stimulated, whether by implants or by ketamine—in other words, the kind of closed-loop device neuroscientist Helen Mayberg was after. Such a device would be able to listen to the brain's neural code during and after zapping or other stimulation, so you knew what had happened in the brain as a consequence.

A memory chip may not be on the cards, but the neural code has yielded some dividends for Ian Burkhardt. In 2020, Battelle researchers were able to use his existing implant to detect residual signals from his sensory nerves and thereby restore an approximation of sensory feedback.[74] "It's huge because I can know that I'm not going to drop something when I'm using the system," Burkhardt told a reporter from MathWorks.[75]

The future of brain chips

So when can you expect to buy your own exocortex? Certainly not with today's implants. All the experiments I've described so far have been done with a single type of implant: the Utah array—whose design has remained largely unchanged since its invention. Until recently it was the only FDA-approved device for those looking to read or write the neural code. Just to be clear: it's approved for research. Not for you and me. Regulatory hurdles kept more advanced implants from advancing enough to crack the language of the brain. Many devices have produced intriguing results in rats (or sometimes monkeys), but went no further. In the early 2020s, the FDA cleared two new designs for early human trials. For most research, however, the Utah array remains the only game in town.

But as a brain–computer interface, this postage stamp–sized miniature pincushion leaves much to be desired. It can only read signals from at most a couple of hundred neurons, and only the kind that populate the topmost millimeter of the brain. And no, you can't just fill the brain with a whole bunch of them. Any more than a couple, and the wires that connect the chip to the signal processing apparatus outside the skull pose an increasingly serious infection risk. Not to mention the prohibitive amount of data this would generate—more than can be reasonably stored by today's machines.[76]

While dramatic implants like BrainGate got a lot of publicity, after the TV lights and cameras went away, some participants found their devices stopped working. After years spent learning to use her implant with a robot arm, Jan Scheuermann, another tetraplegic volunteer in Donoghue's experiments at BrainGate, gradually lost her dexterity, which felt an awful lot like descending into paralysis for the second time. As the researchers explained to her, this was due to a predictable immune response.[77] It shouldn't come as a surprise that the brain experiences a metal pincushion as a foreign invader and mounts a spirited defense, walling off the implant in a protective sheath. As a result, the Utah array is unlikely to be the foundation of any future brain chip.

There was once a design that ostensibly addressed this problem—in the 1990s, a neuroscientist named Phil Kennedy designed an alternative to the Utah array. Instead of 100 pins sticking their noses into the neurons to eavesdrop on their conversation, his "neurotrophic electrode" worked on the opposite principle—make them come to you. The electrode was a glass cone that housed a gold wire impregnated with growth factors and other tempting treats for neurons. Instead of mounting an immune response, they would grow into and entwine themselves with the electrode, which in theory should keep the electrode functioning for years. Oh, and it was wireless.

In 1998, Kennedy put one of these into a Vietnam veteran named Johnny Ray, who had been rendered unable to move or speak by a stroke; he was fully aware but locked in. Kennedy's electrode was able to pick up Johnny Ray's brain signals well enough to allow him to move a cursor across a keyboard to slowly put together words. The media compared Kennedy to Alexander Graham Bell, but the accolades didn't last. After another one or two locked-in subjects didn't respond as well, and Kennedy couldn't find new volunteers, the FDA rescinded approval for the neurotrophic electrode for human volunteers. Kennedy wouldn't provide clear data on what kinds of things he was putting in the electrodes he was implanting into his volunteers. "It was proprietary," he explains. In a last effort to get enough data to convince the FDA to re-approve his implant, he chose the only patient he could. In 2014, Kennedy flew to Belize to have his own (banned for human use) electrode implanted into his own (perfectly healthy) brain by an increasingly nervous neurosurgeon, for $30,000. The procedure would have been illegal in the US.

Kennedy survived the eleven-and-a-half-hour surgery, and despite a few scary post-operative days during which his state resembled the locked-in patients he used to treat, a few years later he seems to have come through the experience largely intact. Unfortunately, the electrode only stayed in his brain for a few months before there was a problem. A second surgery excised the recording and transmission equipment, but not the electrodes—they were in too deep to safely get out.[78]

The episode cannot have made the FDA want another look at Kennedy's paperwork. Kennedy insists he got good enough data to inform future papers, and that the implants left no lingering effects. However, now there are several other new designs in the works—another electrode called Neuropixels is already being used to record data in patients undergoing DBS implantation.[79] It's not yet approved but has a similar design to Kennedy's neurotrophic electrode, able to record deeper in the brain. More ambitious designs abound, with exotic names like neural dust—micron-size piezoelectric sensors that would be scattered throughout the brain and use reflected sound waves to capture electrical discharges from nearby neurons[80]—and neurograins; salt grain–sized sprinkles unveiled in 2021 to make a better ECoG.[81] BlackRock, the investment fund that financed the neurograins, is on record as stating that they want brain chips to become more common than pacemakers.[82] That is also Elon Musk's stated goal. And in 2023 the

FDA cleared Neuralink for early human trials. The N1 implant, whose self-contained design promises tenfold better fidelity and no wires trailing out of the skull, was sewn into one volunteer in early 2024.[83]

Musk has stated that he hopes his implants will become common ways to control human frailties and help us interface seamlessly with technology. However, as with Kernel's ill-fated foray into memory modification, his vision will require not only reading the brain's signals but writing to them. And I hope it is clear, after reading this chapter, how much harder it is to write to the brain than to read its signals. This is because we don't actually understand enough about the brain to change how it works. "What gets forgotten in a lot of these conversations is how little we know about the brain," says Flavio Frohlich, a neuroscientist at the University of North Carolina. "There are very few facts that have been independently verified—I mean like basic fundamental stuff, including visual processing." This is where the revamped Kernel device to may help. It's new brain-reading headgear—no brain surgery required—that combines the best of both EEG and fMRI: magneto-encephalography (MEG). This provides a kind of Google street view for your brain, showing where electrical activity is happening. MEG used to be feasible only with superconductors, which have to be cooled with liquid nitrogen, so the machine was about the size of one of Einthoven's early ECG contraptions. Kernel's design uses laser cooling; the only remaining problem is that like the early galvanometers Nobili was working to improve, the MEG is overwhelmed by the Earth's magnetic field. "It is wildly bigger than the magnetic field of your brain," says Marblestone. Which is why for now, the helmet looks like a large white plastic mushroom, *Mario Kart* by way of *Spaceballs*. But at least it's useful.

There's a lot of open road between current brain implants and any Silicon Valley applications. Complex and subjective functions like memory would require precisely overwriting the activity of millions of neurons. There is also the problem of how many pins you can actually stick into someone's brain before the brain begins to fight back. This all sounds abstract until you think about the fate of people like Ian Burkhardt, whose paralysis is only temporarily abated every time he volunteers to be experimented on in the lab.[84] Jan Scheuermann, the woman who gradually lost her ability to use her robotic arm, told *MIT Technology Review* reporter Antonio Regalado that she once had her care aides put rat ears and a tail on her in a darkly funny nod to how she sometimes thinks scientists might perceive her.[85] Further advances in brain implants will require many more people like Burkhardt and Scheuermann.

Decoding brain signals will continue to yield phenomenal science, but to become as commercially successful as a Fitbit—another goal Musk has articulated—a brain implant will need to do more than read. It will need to write. But it will be a long time before regulatory agencies allow the necessary trials on large groups of human volunteers. And yet, these are not the messages that dominate about the future of brain implants. Few people feel empowered to call bullshit on the more extreme claims because few areas contain subject matter more opaque and that requires more cross-disciplinary knowledge than neuroengineering. Kernel's story is a rare exception in that the company followed the science even when it did not support the initial goal. Future developers, take note.

And these challenges don't just apply to the brain.

In the media, the brain's electricity gets all the attention. However, the neural code is only one aspect of a much larger system of body-wide bioelectrical signaling.

CHAPTER 6

The healing spark:
The mystery of spinal regeneration

In 2007, Brandon Ingram reached out for his walker, cantilevered himself up from his wheelchair, and, once he had righted himself, began to take small steps across the carpeted living room. It took great effort and some help, but he was ultimately able to control his own legs by using his abdominal muscles.[1]

This should not have been possible. Five years earlier, Ingram had been thrown from a car in a highway accident and the injury to his spinal cord was final. Doctors told him that he would not walk again.

And yet here he was, walking. It was a bit of a technicality, as he still needed a wheelchair for most things, but he had regained other capacities that are far more important when you have a spinal injury: shifting his position, and feeling some sensation. "I'm very fortunate," Ingram told the *Boston Globe*.[2]

His good fortune was that his accident took place just as a new clinical trial was recruiting spinal injury patients at Purdue University in Indiana. Several days after his injury, a neurosurgeon placed electrodes between the vertebrae of his crushed spine, where they emitted an electric field. This field, the researchers hoped, would coax opposing ends of the severed motor and sensory nerves in Ingram's spine to slowly crawl toward each other over the damaged site and knit themselves back together, allowing the brain's signals to flow through once more unimpeded. The implant was removed after a few months. When the researchers followed up with Ingram and his cohort a year later, most trial participants reported some improvements.

In 2019—twelve years after Ingram's tentative steps—the scientist who invented the device that mitigated Ingram's grim prognosis died, and with him, much of the expertise around the oscillating field stimulator. Though it had been deemed safe and appeared to do what no other drug or technology had ever managed, and a larger trial had already been tentatively approved by the US regulatory authorities, soon after Ingram's comments to the *Globe*, the research stopped cold.[3] Only fourteen people have ever benefited from it, and after years of having its development blocked at every turn, the company tasked with bringing it into the world went bankrupt. The device was mothballed.

There are still people who get very upset about the circumstances around its demise. "I think this set back spinal injury research by ten years," says James Cavuoto, who runs the influential industry publication *Neurotech Reports*. "Where would we be today if they hadn't scared off the researchers and investors who wanted to pursue this line of research?" Sidelined by an establishment that didn't understand the principles behind its function, attacked by point-scoring competitors whose animus was more personal than professional, the oscillating field stimulator, Cavuoto reckons, was simply too far ahead of its time and too unfamiliar to succeed—too much of a departure from how people were thinking about interfacing biology and electricity at the time.

That's because this implant wasn't targeting action potentials: it aimed to harness a more fundamental electrical field, whose existence hadn't been officially acknowledged until the 1970s. The same electrical signature is emitted and used by skin, bones, eyes, by every organ in your body. New research and tools have begun to shed light on the physiological underpinnings of this bioelectric field, illuminating its inner workings and medical potential. The 2020s will bring more devices and techniques designed to manipulate it. As usual, though, to understand this tale in full, we have to go back (very briefly!) to the beginning.

Lionel Jaffe's lab

It all started with those old studies that showed everything produces its own electric field—brainless organisms like hydroids, algae, oat seedlings. Lionel Jaffe began trying to unpick the mystery in the 1960s, when few electrophysiologists wanted to study anything but the nervous system. But Jaffe, a Harvard-trained botanist with the soul of a physicist, was in pursuit of bigger, more unifying theories.

A good place to start was with brown algae (that's marine fucus to you, if you're nasty). Fun fact: fucus contains up to eight times more sodium than cheddar cheese and eleven times more potassium than bananas. Maybe we'll all be eating it in the future. But the reason biologists love the stuff is that it's a sexual organism that discharges its sperm and eggs directly into seawater (you're welcome). That makes it possible to study its entire developmental procession from day one, without having to navigate any tricky uteruses. The algae grow differently on one side versus the other, based on which end of them is exposed to sunlight.

To investigate their electrical characteristics close up, Jaffe got a bunch of the fucus fronds and put them into a Purdue hot tub to commingle their efflux. Once he had his new developing embryos, he put them into a neat row in a narrow tube, shone some light on one end to mimic sunshine, and checked whether there was any measurable electrical field as the embryos began to grow. And boy was there. Positive up, negative down. It was like a battery. Now he needed some smart kids to help him investigate why.

Purdue was one of the premier electrophysiology institutions in the world, so talent was plentiful. Jaffe decided to poach the most promising students from the physics department. His first catch was Ken Robinson, who abandoned vacuum physics after attending his first Jaffe class. He was in awe. "He had the most accurate and intuitive understanding of physics and mathematics of anybody that I ever knew," Robinson told me. "I was just blown away."

Next, Jaffe turned Richard Nuccitelli's head away from solid state engineering. "Who ever would have thought cells could make electrical currents?" he marvels, fifty years on. He left physics and crammed entire semesters of biology to try to catch up with the rest of his labmates. They were glad to have him. "He was the most gifted technician I have ever laid eyes on," said Robinson. In 1974, Nuccitelli built Jaffe a brand-new electrical measurement device called a vibrating probe. It was a hundred times more sensitive and powerful than anything that had come before. With this, the newly assembled team could start to properly investigate the minuscule electrical currents that swirled around the surface of fertilized fucus eggs. These currents were much, much smaller than the ones that make an action potential go pop. The team christened them "physiological currents." In addition to being weak, they were also steady: where action potentials oscillate like strobe lights, the physiological currents shone from the organisms, steady as a lightbulb.

The field seemed to orient the fucus, enabling it to grow correctly toward the sun. What would it do for other creatures?

Jaffe decided to make his own weak electric field, being careful to exactly mimic the faint strength of the physiological fields that emitted from brown algae eggs naturally, and apply it to other living things.

First up: the muscle cells of frogs. These had been chosen by Robinson and Mu-ming Poo, the biophysicist on the team, in keeping with the centuries-old tradition that had been inaugurated with Galvani's preparation. They placed the cells into a petri dish, put them under the physiological field, and watched. A curious behavior emerged. Not only did the neurites (the broad name for the outgrowths of Neurons that includes axons and dendrites) extend more quickly toward the positive electrode; their very molecules were redistributed by the electric field.[4]

To go back to Faraday's ions for a minute—it's not just ions that wander toward their preferred electrical "side." Turns out, whole cells can wander too. Jaffe's group hadn't been the first to observe this phenomenon—people had been seeing electrotaxis (the migration of cells under electricity) since the 1920s.[5] It genuinely freaked people out. There was simply no plausible explanation for a bunch of cells crawling around a petri dish chasing an electric field. People chalked it up to a poorly understood chemical effect and did their best to ignore it. What was different now, in Jaffe's lab, was that for the first time, there was an instrument and new knowledge to study this phenomenon properly.

The experiments and theories that came out of Jaffe's lab united a whole branch of cell electrophysiology, which because it took place outside neuroscience, had previously been littered across a collection of separate disciplines. Jaffe's lab was considered a second home by many of his students. He was incredibly dedicated to his science and to the scientists who worked under him. Robinson was inspired by his fearless approach to seeking the truth. "He never let data follow a hypothesis—only the opposite," he told me. "If you got a result that didn't match the data, he didn't get upset about it," said Nuccitelli. "He would drop everything and say 'we have got to pursue that and find out what it's telling us.'" Mu-ming Poo is now one of the towering giants of neuroscience, with joint appointments at the University of California Berkeley and the Chinese Academy of Science. Jaffe's inner sanctum was his lab family. Then Richard Borgens came to town.

The Texan

There were some missing years on Richard Borgens's application to study in Jaffe's lab. Jaffe asked about them. Instead of an answer, Borgens handed him a vinyl record of his band, the Briks.[6]

Borgens was from Texas, and the only thing bigger than his personality was his moustache. He liked vintage cars, vintage guns, and amphibians (from an early age, he had been transfixed by how newts in his father's aquarium regrew their legs after the fish bit them off). His path to Purdue was very different from the high-caliber institutional pipeline that had produced Ken Robinson and Mu-ming Poo.[7] In the late 1960s, he had started an undergraduate degree at the University of North Texas, where he was soon diverted by the Denton music scene (most of his bandmates attended Cooke County Junior College, designated by one teacher a "home for the academically ill"[8]). He was vocals and lead guitar, and the band's moody, melodic twang sufficiently captured the zeitgeist to attract a dedicated fanbase and have a few songs go national. On weekends, Borgens liked to jam with Stevie Ray Vaughan's older brother, or when he wasn't around, Don Henley. "All of us who followed them thought for sure they were headed for greatness," wrote one fan on a memorial site forty years later. "But time, the military, and the war in Viet Nam [*sic*], and the general insanity of the time, all saw it another way."[9]

Borgens ended up on a brief tour as an army medic and came back with a different perspective. He quit the band and finished his degree, completed a masters in biology, and then arrived at Purdue. Borgens recalled walking into Jaffe's lab and seeing people doing things he already knew how to do to get a PhD—"Why couldn't I get a degree doing that kind of thing too?" Borgens was the only one in Jaffe's lab who wasn't a physicist, but that didn't bother either of them.

Like Jaffe, Borgens was too impatient to study parts of a system—he wanted to understand how they worked as a whole. If his approach to academia was slightly less buttoned-up than some of his labmates, he won them over quickly enough. "He came across as a hick, you know, but he was really very smart," says Nuccitelli. The two became fast friends when Borgens found out Nuccitelli played string bass: "We wrote a lot of songs together making fun of the people in our lab."

But mostly they played with what they could get electrical fields to do. Borgens liked to call himself an experimental zoologist.[10] For a while he was diverted by a project in which he tried to use electrical fields to grow legs on a snake. Mu-ming Poo was long gone—he had abandoned electrotaxis in favor of the more obviously explicable, scientifically acceptable mechanism of using chemicals rather than electricity to coax neurites around a petri dish. But the other students in Jaffe's lab retained their commitment to physiological fields even after they left Purdue—Robinson going to Connecticut, Nuccitelli to California, Borgens to a fellowship at Yale. In 1981, Robinson and his student Collin McCaig published definitive proof that the cells in the dish were responding to the electrical field and not any other mysterious chemical signals.[11] They found that it was possible to "turn" the neurites' growth in any direction you liked just by re-orienting the field. This worked so well and so predictably that by continually changing the location of the field source, they were able to "draw" intricate patterns. They made a game of looking for their own initials in the axons' looping scrawls.[12]

As they were starting to discover this power of manipulation, new implications arrived: those same electric currents they had measured with the vibrating probe were also involved in regeneration. They had been shown leaving the cut ends of amputated amphibian limbs, raising the possibility that they were causative agents.[13] Borgens took the petri dish research from Jaffe's lab, and in 1981 extended it into living vertebrates. He started with larval lampreys.[14] What's unique about these sea creatures is their ability to spontaneously regenerate their spinal cord if it gets severed. The process normally takes about four to five months, and during the healing process you can observe those physiological fields and currents driving out of the injury just as clearly as du Bois-Reymond measured his own wound current.

Borgens wanted to know if you could amp them up. When he applied an electric field to the regenerating neurons, he was able to speed up the healing time by a factor of three. The reason this had worked, and that the electric field was able to speed up the spinal healing process, is because it had prevented the severed axons from engaging in a behavior called die-back. Die-back is one of the most vexing obstacles to healing any spinal injury, mammalian or amphibian. When neurons are cut, they initially respond by shrinking away from the cut edge before beginning the process of regrowing. If you could prevent the die-back, you could head off a bunch of the other problems that pile up after a spinal injury.

Dying and injured cells void their toxic internal contents, which inadvertently kills nearby healthy cells. That brings the macrophages and white blood cells, charged with tidying up debris and gobbling up foreign bodies, marching into the injury site. But these cells are not great at portion control; they invariably overeat and overstay their welcome, which creates a big fluid-filled cyst. Then scar tissue begins to form, creating yet another physical obstruction for any axon thinking about regenerating. As if all that weren't bad enough, in adult mammals, injury leaves behind inhibitory molecules that unambiguously signal that something bad has happened here, do not enter. No wonder so few vertebrates can regrow a spinal cord.

Borgens had an idea for how to overcome this. He figured if you could get axons to grow across this territory right away before all the other chaos kicks off, there was a much better chance of regeneration. Poo had already established that neurites grow faster when you put them in a DC electric field, and that they grow toward the cathode. Sure enough: when Borgens applied the electric field, it acted as a coach and guide. It coaxed the axons to ignore the normal inhibitory cues that prevent reconnection to their lost other halves. And this was happening in the complex environment of a living lamprey, not some petri dish.

In 1982, he returned to Purdue, where he quickly took his lamprey finding up the ladder into mammals, suturing electrodes to guinea pigs' severed spinal columns. The experiment yielded the same results: he found that, again, he could trace the regeneration of axons across the lesion site. But he ran into a problem he hadn't seen in lampreys. Healing in the guinea pigs was sporadic and depended on whether the cathode was above or below the injury site.

The spinal cord is organized like a two-lane highway. The sensory axons go up to the brain to deliver sensations. The motor axons come down from the brain to carry instructions. So, if you put the cathode above the injury site, all the axons will grow toward it—meaning only the sensory axons will reconnect across the site. Put the cathode below the site, and only the motor axons cross. However, Borgens remembered that Robinson had previously demonstrated in frogs that neurons grow eight times faster toward the cathode than they do toward the anode. He realized that if he could get his electric field to *strobe* instead of shine steady—reversing its polarity back and forth so the cathode would be on one side of the injury for fifteen minutes, and then switched to the other side—he might be able to solve the problem. To everyone's surprise, it worked: Borgens was able to create a kind of two steps forward, one step back pattern of progress that eventually coaxed all of the axon pieces to fuse together. The guinea pigs regained motor and sensory function.[15] He called his new invention the extraspinal oscillating field stimulator (OFS).

By then, both Borgens and Robinson were back at Purdue, ready to continue the work begun by their mentor Lionel Jaffe—who had meanwhile left the university to direct the brand-new National Vibrating Probe Center at the Woods Hole Marine Biological Lab (and to devote more time to his marine fucus)—but they had different ideas of how to do so. Borgens had already set his eye on medical applications. The implications were so obvious. Human spinal cords don't heal naturally. But if guinea pig spinal neurons could be coaxed to regenerate by an applied electric field, it stood to reason that this technique should help heal the same kinds of devastating injuries in people.

It was a good time to be doing this kind of work. Optimism about spinal injury research was ascendant after a long period in the doldrums, aided by a spate of high-profile injuries. Marc Buoniconti, the son of a Super Bowl–winning Miami Dolphins linebacker, had just been catastrophically injured playing college football. In 1985, his father helped found the Miami Project to

Cure Paralysis.[16] It was one of several influential efforts formed in the US and Canada to address spinal cord injury, all of which attracted considerable funding and media attention. It was one of these organizations that invited Borgens to attend a philanthropic dinner for spinal cord injuries, recalls Debra Bohnert, the lab's administrative director from 1986 to 2018. "He came back and said, 'I can do this, I can figure out how to cause regeneration in the spinal cord,'" she says, "and that's what we ended up doing for the rest of his career." His passion clearly impressed the philanthropists he had met at one of these dinners, because in 1987 one of them, a Canadian millionaire who used a wheelchair, donated a pile of money to Purdue, earmarked for Borgens. He used it to establish the Center for Paralysis Research at Purdue's School of Veterinary Medicine.

With the new influx of cash, and the upgrade to a whole new building, Borgens set his eyes on his next target. He wanted to get the OFS into human trials, but you couldn't go to the FDA with a guinea pig or rat trial—the diameter of their spinal cord is an order of magnitude smaller than that of a human. The effects from an electric field would differ so much as to make the trials meaningless.

So Borgens settled on dogs. It wasn't just the better approximation they offered of human anatomy; he was also interested in them because they offered a chance to treat real-world injuries. When dogs sustain spinal injuries, the injuries tend to have a lot in common with the kinds of spinal damage people face—messy crush injuries, not artificially tidy scalpel incisions made in a lab. This could be the springboard to a human trial. (Also, it must be said, Richard Borgens loved dogs.)

So he got in touch with a company that makes assistive devices for paralyzed dogs called Doggy Kart. You may have seen dogs wheeling around in these—they look like a child's toy wagon. The dog's back half is strapped in while the dog uses its front legs to pull it along. As cheerful as this situation may look to passersby, for dogs and their owners, paralysis is a grim situation. When a dog is paralyzed, the owner must manually express the animal's bowels and bladder several times a day. A veterinarian will usually recommend euthanasia.

Bohnert says the center offered to pay for a dog's spinal surgery if the owner would agree to let them implant the stimulator. "We also gave them a wheelchair," she says. "We just asked them—hey, if your dog gets better, we want it back." The first trial was on twenty-four dogs, thirteen of which were implanted with a real stimulator.[17] (An important note here is that, at this point, there was no more ability to test whether the neurites had grown the way Borgens was expecting them to. People's pet dogs are not lamprey eels—you can't kill them after a test and dissect their spines to examine how well the neurites responded to the electrical stimulation. All you can do is figure out from the dog's behavior whether there have been any meaningful changes.) After six months, seven of the OFS dogs could walk again, two of them almost as well as a dog that had never been injured. All the rest regained control of bowel and bladder and other functions. The gains were permanent.[18]

Based on this success, in the early 1990s, they expanded the trial. People sent their dogs from all over the country to take part. Borgens recruited Scott Shapiro, a neurosurgeon at Indiana University, to help him implant more OFS devices in more animals. By 1995, they had treated nearly 300 dogs with spinal injuries. "Without treatment, 90 percent of these dogs would have been put to sleep," Borgens told the *Chicago Tribune*.[19] "We got a lot of carts back," says Bohnert.

Saving dogs from paralysis and euthanasia simply had no PR downside. These successes were captivating, and Purdue was awash in the warmth of media attention and money. In 1999, Borgens got a provision written into Indiana law that committed the state to give half a million dollars to Purdue each year for spinal injury research.[20] The following year, Mari Hulman George, the chair of the Indianapolis Motor Speedway—if you watched the Indy 500 between 1997 and 2015, it was her voice you heard bellowing "Ladies and gentlemen, start your engines!"—added another $2.7 million to the pot.[21] There was now enough money to pursue a human trial. Shapiro and Borgens started the long process of getting approval from the FDA. "It took two years and four volumes of text but we got approval to put ten devices in ten patients," says Shapiro.

Purdue held a grand, official announcement to kick off the human trial. Wandering around the stage was a shiny brown pointer mix called Yukon whose life had been saved four years earlier by Borgens and his team after being paralyzed by a ruptured disc.[22] David Geisler, whose family Yukon was part of, recounted the harrowing circumstances around bringing his beloved pet to the center for evaluation to see if he qualified to join the trial, knowing the outcome if the answer was no. "There was a team of people watching me cry," he said. But the OFS came through. "I knew he was getting better when he started wagging his tail," said Geisler.[23] By the time the human trial was announced, Yukon was bounding up and down the stairs again. Hope was thick in the air at the press conference. The event was covered by the *Los Angeles Times*.[24] The bar had been set high.

Brandon Ingram and the nine other volunteers had been paralyzed during a narrow window less than twenty-one days before starting the treatment. All their injuries had been catastrophic. Borgens and Shapiro implanted the device about the size of a cardiac pacemaker and left it there for fifteen weeks. In that time, they hoped, its oscillations would guide the axons across the injury the same way as they had been shown to do in lampreys, rats, and guinea pigs. And they hoped for the same functional and sensory results they had seen in the dogs.

After the devices were taken out, Borgens and Shapiro then followed the participants for one year, testing them periodically for what sorts of changes they perceived. Unfortunately, few reported changes in mobility like Ingram experienced—but that is far from the only endpoint for spinal surgery. In surveys, people with spinal injury consistently put regaining the ability to walk at the bottom of a long list of much more pressing concerns, which include going to the bathroom independently, regaining sensation, and the ability to subtly shift position to prevent pressure ulcers. The volunteers regained a mix of these abilities.

After a year, all but one of the participants had regained enough sensation to feel their hands and legs. It was mainly light touch and pain, sexual function, and some proprioception (the body's sense of its own position). No one regained bowel and bladder function. This wasn't a disappointment, as Borgens never claimed that he'd make people walk again: "Richard was very careful to say to us: '*never* say we're going to cure paralysis—just that we will give some people some function back,'" Bohnert told me. Two patients—including Ingram—did recover some lower-extremity function, and both recall that another of the patients could now lift his legs horizontal to the floor for the first time since his injury. Most importantly, any ability that had been regained was there to stay: "Their improvements were durable," Shapiro says.

The results were impressive enough that the editors of the *Journal of Neurosurgery: Spine* put them on the cover in 2005. This was a phase one clinical trial, which is only for safety, not for efficacy, so the functional improvements "didn't count." But that was fine—the device had cleared its first hurdle: no deaths, no infections, and no painful side effects. The OFS had been proven safe.

Eventually, it would have to pass several more trials in order to be sold. A device like this is not permitted to be sold on the open market in the US without explicit regulation by the Food and Drug Administration. If the FDA doesn't clear it for use in human beings, you can't sell it. Period.

Of course, none of the subsequent headlines screamed about the device having passed a routine safety clearance. "Nerve Repair Innovation Gives Man Hope" was more like it.[25] Ingram's experience was especially galvanizing. He was able to dress himself, shower, and get in a car on his own, he told newspapers, which continued to seek him out two years later.[26]

Based on the safety analysis of the first trial, the FDA approved a second clinical trial for ten more patients with severe spinal cord injuries.[27] This one would differ from the previous trial in a crucial way: instead of simply ensuring the device was safe, it would explore how well it actually worked. The most important way to do that in any scientific study is by equipping some people with the real device and others with a sham—a pseudo-stimulator that doesn't do anything. This placebo control group is the key to any "gold standard" trial. They provide the crucial contrast to your technique. If there's a big difference between their improvement and the better improvement of the people with the real device, you move on to bigger and bigger trials to determine, to ever greater precision, whether the effect of your device is real and how big it is.

Neurosurgeon Scott Shapiro methodically went about planning all of these next steps. He recruited three more neurosurgeons at other medical centers that had agreed to take part in the next small randomized controlled trial. And after that, his plan was to approach the NIH to fund a slightly bigger trial of eighty patients, forty of whom would receive a functional OFS. To Shapiro, the next steps were clear, and they had to be done in order.

That wasn't how Borgens saw it. He was getting older, and he had been carrying out this work now for the better part of twenty-five years. He was tired of baby steps, and all the publicity and acclaim was making it harder to think in increments. He wanted something on the market. But the OFS was a hard sell for big device manufacturers like Medtronic. There was no money to be made on such a comparatively rare injury. What was even more unpersuasive to them was that most of these injuries occurred in uninsured males (more likely to sustain gunshot and diving injuries). Borgens thought he could manufacture and sell the device to a big, rich company and let them worry about all the FDA paperwork. So, three months after the paper was published in the journal, he and some of his colleagues founded a start-up called Andara Life Sciences and negotiated the intellectual property rights to the OFS. Within a year, they had found their big, rich company, and it quickly swooped in and gobbled up Andara.[28] The company was Cyberkinetics, the same outfit that had made BrainGate. Yeah, those guys.

Purdue bathed in the golden light of Borgens's fame. His research was raining both state and private money on the school. Mari Hulman George reached into her foundation's pockets again for another $6 million. In short succession, the OFS won a series of major accolades from industry watchers who eagerly anticipated the day when this first-ever neural regeneration device would hit the market: it "represents a groundbreaking advance in neurotechnology," pronounced James Cavuoto, the editor of *Neurotech Business Report*.[29] As an investigational device, however, the OFS was still not approved for sale, and was available only through participation in a clinical study (the details of which Shapiro was still meticulously working through). Cyberkinetics, which wanted to start earning money from its latest investment a lot sooner than that, filed an application for Humanitarian Use

Device status from the FDA, which, if granted, would have let Cyberkinetics sell the device commercially by late 2007. Borgens and Bohnert were given to understand that the approval was almost a formality. "They told us: oh yeah don't worry about that, we know how to deal with the FDA," says Bohnert. Cyberkinetics expected that they would start selling the stimulators the following year.

The apostate

Ken Robinson was worried. He had wondered, reading Borgens and Shapiro's paper, what accounted for the decision to choose exactly fifteen minutes between switching between cathode and anode. But when he poked around in the references for the justification—he found himself faced with his own name. This was a big surprise for Robinson, as the paper in question had in no way addressed this idea. "It misrepresented my work," he says.

Robinson had never seen any mammalian neurons respond to physiological fields the way his amphibian neurons did. To get any effects at all required one or two orders of magnitude higher fields. So Robinson tried to repeat the experiment in zebrafish. It should have been a formality, a box-ticking exercise. Instead, his zebrafish neurons were unmoved by any interaction with the "physiologic" currents. "We were stopped cold," he says. "You could not extrapolate from amphibians to other animals and assume they were the same, especially mammals. And that caused me to start looking at the whole picture."

In 2007, Robinson brought his misgivings to Shapiro in a long letter, asking whether the team had ever actually directly observed the vaunted bidirectional growth in any of the frogs. "Well, I got no answer whatsoever," he says. Robinson began to worry that the experiments had been unethical. "They just did not have sufficient grounds for doing it."

How could a scientist make any claims for an intervention when there was no one in the study who was not receiving the treatment, as was the case for this safety trial, by design? How could they account for the placebo effect? There was no way to check the volunteers' self-reported improvements against spinal neurite growth—it's not like you could cut them up for dissection. The comparison groups Shapiro and Borgens used were case studies from other experiments unrelated to theirs. Sure, none of the ten subjects were harmed, but "those experiments were unethical even if they didn't harm anyone," Robinson insists. Skipping the necessary foundational work, he says, meant the design of the stimulator was utterly arbitrary, and this alone rendered the experiments unethical. Founding a company and selling the device compounded the offense.

This argument formed the rough gist of a review article, which Robinson and his colleague Peter Cormie published in 2007, after deeming that they had waited long enough for a response to their letter.[30] It eviscerated Borgens's work. Its most immediate effect was to alienate Robinson from the rest of Jaffe's descendants. The break was so immediate and complete that today, retired in Oregon, he still refers to himself as "the apostate." This term is usually reserved for religious acolytes who turn their back on their beliefs. The acerbic review was the first drop in a hailstorm.

Against expectations, the acquisition by Cyberkinetics had not cleared a path for trials. In fact, nobody told Shapiro about the acquisition. "I was unaware," he says. He had been plugging away at his fastidious roadmap, having implanted another two patients, and things were going well. "All of a sudden, this company came in, took all my study documents and devices, and shut me down."

What no one knew was that, by 2007, Cyberkinetics was near

bankruptcy and desperate for a commercial product. "They tried to ram the device through the FDA on a compassionate use basis on the basis of twelve patients," Shapiro says. "I knew that would fail." At some point—in addition to failing to grant the humanitarian exemption—the FDA also appears to have withdrawn its approval for any more implanted patients in a phase two trial.

But none of this was ever made clear to Borgens. The FDA just dragged its feet on approvals, in a move that seemed calculated to run out the clock until no one had any money left, and the project withered and died.

Quick aside for a moment in defense of the FDA: the Food and Drug Administration might be the most underfunded, overworked, and unfairly maligned regulatory agency in the US. They are tasked with making sure every drug and device meets its claims and doesn't kill people. Business-friendly political administrations like to starve it of funds, believing that the agency really enjoys getting in the way of innovation. But when the FDA can't do its job properly, you get things like vaginal mesh implant disasters and breast implants that leak. It's thanks to the FDA that faulty ventilators were recalled during the Covid-19 pandemic before they could kill someone.

However, as the oscillating field stimulator was going through its tribulations, the FDA was very different from the agency it is today, and this played a part in the device's demise.

Like James Cavuoto, Jennifer French was frustrated by the way the agency handled the Andara device's approval. French, a patient advocate for the FDA, saw the entire process from the inside. She knows a thing or two about spinal injuries. In 1998, after a snowboarding accident permanently damaged her spine, she became tetraplegic. A year later, French volunteered to become one of the first people in the world to test a cutting-edge new electrical implant called the Implanted Neural Prosthesis, which temporarily restores to paralyzed people the ability to stand and move. It does so by injecting pulses of current into the muscles and nerves through precisely placed electrodes. Serving as a test pilot for this cutting-edge neuroengineering gave her unparalleled insight into the gap between what people needed and what researchers were providing. She quickly became involved in advocating for people with neurological conditions, specifically to help the agencies tasked with sorting the promising advances from the snake oil.

For French, the OFS's ability to restore sensation had been the most compelling and statistically significant outcome. Sensation is an absolutely vital priority for people with spinal injuries. It is crucial to warding off pressure sores—breaks in the skin. When you can't feel your skin, these can go unnoticed and then become infected and turn septic, poisoning the blood. Sepsis is one of the top two causes of death among people with spinal cord injury.

However, when it came to evaluating evidence of a product's efficacy, the FDA was less interested in what patients in trials had to say about a device's impact on their life than in what it considered more "objective" measures, which is why, in 2007, the measurements the FDA used to evaluate a device didn't include sensation. "It's considered a black box," she explains. The sort of evidence they wanted could be evaluated independently by a clinician with a focus on motor activity. Today, thanks to the advocacy of people like Jen French, that has changed, and the FDA takes patient-reported outcomes much more seriously than they used to.

But at the time, this meant that the FDA didn't really see what the fuss was about. This device wasn't really getting people on their feet again, so why rush through more trials or exemptions? The other issue was that, unlike today, the agency hadn't yet established the programs that help guide companies through the reams of paperwork necessary to gather the safety and efficacy data needed for approval. They just sort of let them figure it out for themselves. Some did; others didn't.

All the while, Cyberkinetics continued to bank on the exemption, which kept being promised and delayed, month after month. "Back then the FDA was taking a long time to make decisions," says Cavuoto. Even as they passed over patient reports, they attended to reams of other evidence in their deliberations. That probably included Robinson's review. But it would have certainly included the public statements made by a world-famous neuroscientist in a leading newspaper. In 2007, Miguel Nicolelis at Duke was interviewed about the OFS by the *Boston Globe*.[31] "I have nothing good to say about this company," he fumed. "I see no solid science behind their latest attempt to make some quick revenue or save their stock price from collapsing altogether."

What did Miguel Nicolelis know about physiological fields? Not much, it turns out. "This had nothing to do with Andara," says Cavuoto—it had everything to do with John Donoghue, one of the founders of Cyberkinetics. Nicolelis hated Donoghue, he recalls. Both men were pioneers in brain–computer interfaces, but Donoghue was the media darling who got column inches in *The New York Times*, and Nicolelis wasn't—to his ever-present consternation. "Nicolelis didn't know anything at all about the technology—all he knew was that it was John Donoghue's company."

After the disparaging comments were published, Cavuoto wrote a pleading editorial—aimed at the FDA—begging them to not listen to Nicolelis. But it was too late. Cavuoto reckons the article helped sink the company. "The FDA took their sweet time and [Cyberkinetics] ran out of money, and the investors pulled the plug. And that was that," he says. Then the 2008 recession hit. Cyberkinetics, Andara, the humanitarian device exemption—all of it was gone.

Fifteen years later, Cavuoto is still sore about the way things went down, and not just about the device itself. "When you forced

Cyberkinetics out of business—which is what the FDA did—you were sending a clear message to the research and the investment community," he says: work on this, and your career will go nowhere. The effects of its actions, he claims, ripple out more than a decade later. "In my opinion, it set the field back maybe ten years."

The end of the road

Meanwhile, the Purdue lab could never get their phase two trial properly underway. "The FDA wouldn't allow it to start," says Bohnert. "They just kept requesting more information. We never knew why. I just remember saying to Richard, 'Who did you piss off?'" But true to his personality, after all the setbacks, Borgens still refused to give up. He tried several times to reassemble what pieces they had left. Both he and Shapiro—and later, his post-doc Jianming Li—applied herculean efforts to get the OFS restarted or at least prevent it from being memory-holed. In 2012, Shapiro wrote a post-review publication to report the OFS results that included the four additional participants he had managed to cobble together. He published another similar review in a European journal in 2014, part of the same effort to keep the work academically relevant.[32]

But eventually, it was too much even for Borgens. "The FDA just buried him under so much paperwork that he eventually gave up," says Ann Rajnicek, who earned her PhD under Robinson but cut ties after the events. "He used to put his hand up, you know, extend his arm as high as it can go, and he would say, 'I have filled in paperwork, literally this high for the FDA, to try and make this happen.' And he said, 'I just don't have the will for it any more.'"

Li, now a research professor, tried to take over. He modernized the OFS electronics and tinkered with optimizing the electrode placements. The technological advances since 2001 offered staggering upgrades: the ability to alter device settings, new algorithms, control of the device using an app. But Borgens had already turned his focus away from technology and toward drugs that could fuse neurons.[33]

Then, in 2018, he was diagnosed with prostate cancer. That's when Purdue cleaned house. Bohnert says one of the deans forced Borgens to retire. The others in the department who weren't laid off either retired early or left. Borgens passed away at the end of 2019.

Even then, Li tried to carry on his old mentor's work, trying to keep both the Center for Paralysis Research and the OFS alive.[34] As the original patent had expired, Li put together a new patent filing and published some of the advancements.[35] A tentative collaboration with Case Western was on the verge of testing the new version of OFS in humans. Then Covid struck.

In the chaos, Li was let go and replaced with a new director, who changed the mission of the center in a direction that ruled out further work on the OFS. "It was just so sad," says Bohnert. "He had a way of helping people and they wouldn't let it go forward." Shapiro retired from Indiana University in 2021. All that remains of Richard Borgens at Purdue is an office door painted to look like the Texas flag.

Decades later, Andara is a whole garden of what-ifs. Would it have succeeded if it hadn't gotten caught up with Cyberkinetics, hadn't been torpedoed by the recession? Was Robinson right that Borgens skipped crucial steps at the beginning that would have made themselves known in later trials? Or was the idea simply too far ahead of its time?

Here was a device that worked on neurons but had nothing to do with familiar concepts like the neural code and action potentials. It was a wound-healing mechanism that used electricity, which didn't make any sense to biologists and dredged up old memories of electroquackery. "The stuff he was doing was truly on the cutting edge," says Richard Nuccitelli. "Trying to regenerate the spinal cord—guide new growth—that's something standard electrophysiologists know nothing about. They aren't interested in that—they're only interested in action potentials."

Today, spinal stimulation is back in the news.[36] But it's action potentials that are the focus of these new research efforts, which approach spinal connectivity from the traditional perspective. Instead of trying to reconnect severed axons, they apply intense bursts of electricity to any remaining intact axons in the spinal cord to force them to carry action potentials that drive motor function. These few remaining undamaged pathways, it turns out, can exhibit plasticity like that most commonly associated with the brain. This—not fusing broken neurons back together—seems to be how electricity is popularly used in spinal injury research now. The results are dramatic. A handful of people are walking who, before technological intervention, did not. "Maybe if Andara had been approved, that all would have happened years earlier with more different approaches," Cavuoto grouses. "Maybe more people would be walking around now after spinal injuries."

Was Borgens's device actually working on these principles? Or are today's devices actually working because some of the fields are reconnecting axons the way the OFS did? The problem is that there are not a lot of ways to test whether Borgens was right about exactly what mechanism restored movement to Brandon Ingram. As with dogs, you can't cut people open to check.

However, the work emerging from other areas of bioelectricity in the body could settle that question soon. That's because it's becoming clear that Richard Borgens was tapping into bioelectric properties of cells that are just now beginning to be understood in full. When it comes to the physiological fields Borgens's OFS was recruiting, those were absolutely real, and they are not unique to cells in the spine. The same electrical properties are common to every living cell in your body. Borgens might have recruited them in a way that put the cart before the horse, but it is clear that he was tapping into something elemental. And as this research finally begins to mature, it is cohering into theories of how physiological electric fields mend the body whichever way it breaks down, and how to create new devices to help it do better.

Borgens's work continues to be replicated in small trials, most recently by a Slovakian group in 2018, which re-created the OFS precisely. They tested it in rats, and with the improved imaging and analysis available more than thirty years after Borgens did his own rat trials, were able to see exactly what the OFS was doing. Under the guidance of the electric field, the mangled axons successfully clasped on to their estranged partners over the injury site. We may not have seen the last of the oscillating field stimulator.

Borgens's instincts, it turns out, had been right.

All your batteries

In the decades Borgens spent fighting his battle, other researchers rapidly populated the periodic table of all the other cells that respond to ultra-weak physiological electric fields.

Colin McCaig set out to build an unassailable body of evidence that nerves and muscles aligned themselves under a weak electrical field. He realized he needed to bolster his case for skeptics, and that he could do so by showing that the so-called physiological field did the same thing in other kinds of bodily tissue. He recruited Ann Rajnicek—Robinson's estranged protégé—and Min Zhao, who had studied with China's top trauma surgeon, to move to Scotland

and join his lab at the University of Aberdeen. Together, they set out to demonstrate that bioelectricity had profound effects everywhere in the body. What else could be dragged around by a cathode?

Pretty much anything, as it turned out. The same subtle fields Borgens had tried to recruit to heal wounded axons—and which Poo had found guiding those spinal neurites—also coaxed crawling behaviors out of skin cells, immune cells, macrophages, bone cells, and just about anything else they got their hands on.

Zhao in particular was shocked by the sheer power these electric fields could exert. Arriving at McCaig's lab, he had expected a predictable series of events to unfold: as usual in science, he would put in some time characterizing yet another interesting factor among so many others, in yet another complex biological process. Sure, the work would be "important"—but, he suspected, not thrilling or actually that consequential. It wasn't going to change the world. That's how it normally is in biology; there are too many factors involved to ever neatly pinpoint the overarching importance of a single one. This was especially true of wound healing, a farrago of interlocking growth factors, cytokines, and other contenders: "Everyone has their favorite molecule and they can show that it plays a significant role," he says. But when Zhao turned on the electricity for a healing experiment, the results blew them all out of the water.

Zhao was stunned. A tiny electric field held veto power over the influence of any other growth factor or gene or anything else people had previously assumed to account for wound healing.[37] The cells did what the electrical fields told them to do, no matter what else competed for their attention.[38] This is the hallmark of an epigenetic variable. "That was when I realized we were working on something far more important than other people, even myself, had expected," Zhao told me.

To his consternation (and McCaig's and Rajnicek's), no one else was interested in their findings. As evidently disruptive as their work was—for better tissue repair, for understanding of embryonic development, you name it—among other electrophysiologists it went largely ignored.[39] *Electricity doesn't do that.* Many scientists looked at it with the distaste normally reserved for homeopathy.

The Aberdeen dream team, however, was undaunted. They pressed on. They had only seen the first glimpses of why these fields were important. The individual cells shuffling around their petri dishes weren't the main point. After all, your body isn't made of a bunch of individual cells milling about, it's made of huge assemblies of them organized into cooperating tissues and organs. They form four main types of tissue: apart from nervous and muscle tissue, there's also connective tissue and epithelial tissue (skin). And the Aberdeen research promised to answer the long-standing mystery of why electricity poured from these when they sustained damage.

Your skin is a tightly coordinated collective of billions of cells. It's organized into three layers of tissue called epithelium, the outside-facing of which is called the epidermis. If you'll permit an oversimplified metaphor, you can think of your skin as a scaled-up cell membrane, but for your whole body. That's especially true from an electrical perspective.

Epithelium generates a voltage across itself. You could interpret it as an "all systems nominal" signal. When your skin is intact, it generates an electric potential so that the outer skin surface is always negative with respect to the inner skin layers.

Where it gets really interesting, though, is what happens when you cut that skin. You sever the epithelial layers of the epidermis, and when that happens, all those sodium and potassium ions, which had been traveling nicely through the neat pathways offered by their gap junctions, leak haphazardly all over the place. If this were a wire you had cut, you'd be short-circuiting it, meaning electricity would flow in every direction. The neat avenues for the current are gone or smashed-up, and so the ions just pour out into every available space.

As I mentioned in the introduction, this is the wound current you can feel if you bite the inside of your cheek and then touch the bite mark with your tongue. The tingle is you sensing the voltage. Ken Robinson used to do a much more dramatic demonstration for his students at Purdue, recalls Rajnicek. He would take an ammeter and project its dial, resting placidly at 0, on a screen at the front of the lecture hall. Then, with a flourish, he would show two beakers of salt solution connected to the meter, and dip his fingers into the solution, to show that it left the dial unperturbed. For his next step, "which I don't recommend doing today," Rajnicek says, Robinson pulled out a razor blade and nicked his finger, and then dipped the bloody appendage into the beaker once more. The needle would swoop. "You could just see the current going right up," she says. "It got gasps from the audience every time."

All that leaking current creates a field whose influence can be felt across some distance within the body. This acts like a combination burglar alarm, compass, and bat signal for surrounding cells. Just as Mu-ming Poo and Ann Rajnicek used artificially generated electric fields to drag individual cells around a petri dish, the naturally occurring field created by the wound current convinces a whole crew of them to migrate to the wound. It guides in and directs the body's emergency workers: the keratinocytes and fibroblasts that rebuild the structure, and the clean-up crew (the macrophages). They all work together to reseal the epidermis. What's even cooler? The electric field directs the cells to the center of the wound. That's your natural cathode—the big red bulls-eye toward which all the migrating helper cells in the body are marshaled.

This starts the repair process. And as the repairs get underway, the wound current and its associated electric field begin to fade out. By the time the wound is healed over, there is no more wound current to detect. This is how it works in all epithelial cells.

And guess what: your skin is not your only epithelium.

To simplify things even further, think of your skin epithelium as the electric shrinkwrap around your body that keeps your insides inside you and your outsides outside you. And just as your whole body is surrounded by the multilayered electrical epithelium you call your skin, so are all of your organs bound by their own individual electric shrinkwrap.

Depending on the organ, the epithelial shrinkwrap is either on the outside or the inside (technically, if it's inside, you call it an endothelium, but it's still the same stuff). Some organs have both—your heart is wrapped in the stuff inside and out. It wraps your kidneys and your liver. It lines your mouth, blood vessels, the hollow parts of every organ such as the lungs, eyes, your urogenital tract, your digestive tract, your vagina, your prostate. I can't stress this enough: it's everywhere. Its main job, just as a cell's membrane creates a boundary that determines what's allowed in and out, is to determine what goes in and out of the organ it wraps (with contributions from the circulatory system). And because both epithelium and endothelium are electric, that means all those things are batteries too. Every organ in your body has a voltage, and it uses that voltage. The reason for the heart battery is easy to conceptualize—the heart literally uses the field to control its beat. "It's an electrical contraction," says Nuccitelli. But you also have a kidney battery. You have a boob battery (the lumen of mammary glands). A prostate battery (looking at you, Alexander von Humboldt). Everywhere current crosses the epithelium, you get a battery.

The eye battery is probably harder to conceive of, but it's the coolest one. The eye has an extra-strong wound current that helps speed up the healing process in the cornea and the lens when they are injured.[40] That's because the retinal epithelium is one of the most electrically active tissues in your body: the reason any of us can see anything at all is because of the electrical currents and fields that eddy about in its many layers, which 1970s researchers christened the "dark current."[41] While it sounds like a homage to Pink Floyd, its name is literal: this current only flows in darkness. Turn on the lights, and the sodium channels snap shut, and a bunch of other signals flip on the color vision.

So: nerve, muscle, and skin, all confirmed electric. That leaves one last category—the connective tissues like bones and blood, which bind and support the others. Are they electric?

Well, you wouldn't be reading a book called *We Are Electric* if they weren't, so I'll spare you the suspense.

Bone is electric too. Bone is a piezoelectric material, which means it's a tissue that can take one form of energy (the compression of running, say) and convert it to another. For example, the stress of your footfalls on your bones makes them grow stronger because the charges bone cells generate in response to this mechanical activity get translated into electric signals that enhance bone growth. Bone also emits strong wound currents when it breaks: voltages appear at fracture sites and help the bone heal its wound.

In short, you can't talk about a living system without recognizing its electrical component. We are nothing without electricity.

So if the body naturally uses its own electricity to heal wounds, what if we could learn how to control it the way we had with the pacemaker and deep brain stimulation?

Playing the field

It was becoming clear that you could disrupt the body's natural repair processes just by interfering with the electrics. Researchers in Scotland found that if they used channel-blocking drugs to inhibit sodium ions, thereby interrupting the electrical signals sent by the wound current in rats, their wounds took longer to heal.[42]

But was the opposite true? Could we also speed up the healing process by amplifying our body's natural electricity? A spate of clinical trials over the past decade increasingly suggests that the answer is yes. Perhaps the most harrowing kinds of wounds are severe bed sores, which can take months to years to heal (if they heal at all) and attack tissue, muscle, and bone deep beneath the skin. Most of the research using electrical stimulation to heal wounds in humans has been done in these kinds of wounds—like deep brain stimulation, this is a method of last resort when nothing else seems to help. After many years of these kinds of experiments, two groups of scientists conducted meta-analyses and concluded that amplifying the natural wound current with electrical stimulation almost doubles their healing rate.

Nor was the effect limited to skin; since the 1980s, a growing body of evidence hinted that the same kind of small electrical currents could accelerate healing in bone fractures, and some suggested it might even help treat osteoporosis.[43] It helps new blood vessels grow into wounds faster and is also beginning to be seriously examined for eyes. Electrical stimulation has even been shown to be effective at aiding skin transplants—it seems to help the new skin take.

There's a catch: the results of all these types of experiments have been broadly positive—but also inconsistent and unpredictable. "The problem here is that it's not optimized," says Mark Messerli, who is working on bioelectric wound dressings at the University of North Dakota. Because we don't understand the mechanism by which the electricity is speeding the healing of the wound, we can't do anything targeted to enhance or improve—or even standardize—the stimulation. And that makes things hard for any doctors hoping to use electrical stimulation on their patients. "To optimize wound healing, we need to understand how it works."

Min Zhao was able to vastly advance this understanding in 2006, when he and the geneticist Josef Penninger undertook the first-ever controlled experiment aimed at pinning down some of the genes that get switched on by electric fields on wounds.[44] This work was reported widely in the news—it finally brought electricity into the legible zone of genes. This was some of the earliest, most robust and tantalizing evidence for the epigenetic power of the electrome.

The next thing to do was find a way to measure the actual electrical field of human wounds. Existing electrotherapy devices apply current with no insight as to what effect it has on a person's own bioelectricity. To change that, you need a device that can help identify whether a person has an abnormal or malfunctioning wound current. No tool had ever been capable of measuring the electric field in the air next to dry mammalian skin—it had always been done on wet frog skin in the controlled conditions of a lab. In 2011, Richard Nuccitelli created a non-invasive device that could deal with human skin, allowing our injury currents to be closely observed. The Dermacorder could sense whatever the nearest voltage is. Hold it up to the skin and it maps the voltage on its surface and correlates it with the depth of the wound.[45] That gives you a topographical, three-dimensional electrical map of the wound. "This was the first tool a physician could actually hold and use on a person," says Rajnicek.

It vastly deepened the understanding of how electricity works in wound healing. Nuccitelli found a strong correlation between the magnitude of the wound's electric field and the progression of the healing—it peaks at injury, slowly decreases as the wound heals, and returns to undetectable when healing is complete. More interesting, though, was the relationship between the strength of a person's wound current and their ability to heal. People with weak injury current healed more slowly than those whose injury current was "louder." Most interesting of all: wound current strength wanes with age, emitting a signal that is only half as strong in people over the age of sixty-five as it is in those under twenty-five.[46]

With better measurements came better experimental results. In 2015, Nuccitelli and Christine Pullar applied electrical stimulation to wounds and, mapping it with the Dermacorder, were able to coax new blood vessels to form, accelerating the healing in all their patients.

Electric healing

The idea of accelerated wound healing seems to be reaching a critical mass. In 2020, DARPA gave Zhao and several researchers $16 million to develop a next-generation wound-healing system. This will not be a sticking plaster like the kind you use when you nick yourself chopping vegetables. The bandage is intended to heal major traumatic wounds, so it will recruit bioelectric healing of multiple kinds of tissues at once—and speed up healing in all of them.

The first proof of concept has been completed: a device that can maintain specific voltage gradients in cells by exerting individual control over ion channels.[47] The other device is a wearable electronic tattoo, circuitry made of electric ink drawn on the epithelium.[48] This traces, in a three-dimensional way, exactly where the wound current travels through the tissue as it heals. Such a bandage is useful both observationally and diagnostically, as it provides something like a topographical map in live tissue. The idea is that you could use it like Google maps to track the exact movements of the wound current's various elements in real time. It can also deliver external electrical current with similar precision.

Rather than just beaming an all-purpose electrical field onto a wound and hoping for the best, it introduces electrical fields in a precise way that guides them to where they are needed.

Zhao reckons that this electrical conductivity body map is similar for all of us, a bit like the wiring in every house adhering to common standards. "You can't just stick your power plug into any random spot on the wall," he says. Richard Borgens was far ahead of his time in trying to harness the radical implications of what Lionel Jaffe had discovered about the body's physiological fields. But in rushing ahead with clinical experiments, he was trying to skip the steps that are only now possible with better understanding of bioelectricity's role in healing and precision tools to map and measure it.

In fact, using a wound-healing perspective may have not been radical enough for what Borgens was trying to do with broken neurons. He was focusing on controlling individual cells. In the past decade, an amazing rush of new research has revealed that you don't have to micromanage at this level—there are ways to switch on the body's dormant control systems to do it all for you.

If you can figure out the right ion channels to switch on and off, you can do a heck of a lot more than heal an injured limb. You could just regrow the whole thing from scratch.

PART 4

Bioelectricity in Birth and Death

> *"We have trillions and trillions and trillions of cells in our body . . . the genes in your nose and the genes in your eyes and the genes in your mouth and the genes in your elbow, meaning the cells in these tissues are all the same. Why do they do so many different things?"*
> Mina Bissell

By the dawn of the twenty-first century, we began to suspect that the signals encoded in all those moving ions do a lot more than patch up injuries. The old view that only neurons send messages that govern communication gradually began to fall away, and a new idea emerged that perhaps all cells send and receive electrical communications. The same physiological fields that guide healing also appeared to guide our body's ability to mold itself from scratch according to a remarkably consistent blueprint—and seemed key to cancer's ability to spread in the body. Understanding this electrical language could offer the keys to life's most fundamental questions and intractable problems—from how we are made to how we are unmade.

CHAPTER 7

In the beginning:
The electricity that builds and rebuilds you

Schrödinger's finger

For the past decade, Michael Levin's conference talks and papers have included a finely detailed line drawing of a little white mouse sitting up on its hind legs. The expression on its face can only be described as a Mona Lisa smile.[1] Another source of ambiguity is its left foreleg, which is encased in a small box. The paw in the box may have five fingers, or it might have four.

There are several actual mice in Levin's lab at Tufts University, and they're each wearing one of the little boxes. All have had a single finger amputated. The box is called a bioreactor, and it is placed on the stump after amputation, along with something patented to manipulate the electrical communications in the remaining tissue. It's possible that one of the boxes contains a once-again complete set of five fingers. The results are not yet in but this Schrödinger's appendage could change the future of an entire scientific field.

"Regenerative medicine" is an umbrella term that was only invented about thirty years ago to cover the wide variety of ways people have tried to replace what has been lost to trauma or age.[2]

The discipline was stitched together, a bit like Frankenstein's monster, from a disparate collection of other subdisciplines that included implant and transplant medicine, prosthetics, and tissue engineering. What united them all into a coherent framework was the discovery, and the galvanizing promise, of stem cells.

The reason you're always hearing about stem cells is because of their unique ability to turn into many other kinds of cells. They're a bit like kids: initially infinitely malleable, but as they mature and grow into their eventual vocation, they specialize into specific adult roles like muscle or nerve or bone. When you were a three-to-five-day-old blastocyst, you were all stem cells (about 150 of them, in fact). By the time you're an adult, you don't have many of them left, and what few you have are mostly generated in your bone marrow.

When, in 1998, it became possible to derive these magic materials from human embryos and transform them into any other cell in a lab, it was suddenly conceivable that we could use them to repair or replace any organ or body part, instead of what we had done before: swapping in metal or plastic, or a donated organ that required suppressing the immune system. Whether they were old, damaged, or diseased, soon stem cells would rejuvenate livers, joints, hearts, kidneys, eyes, and anything else you could want.[3]

Amid controversies (people didn't like the idea of using fetal tissue as building blocks for medicine) and new hope (it turned out other cells in the adult body could be recruited for similar purposes!) the headlines never stopped. Stem cells would cure neurological disorders. They'd cure lower back pain. Heck, they'd cure anything. They were biological miracles.

Despite thirty years of dramatic headlines, however, most of these goals remain perpetually out of reach. "There's no injury, no disease, no anything, where stem cell therapy is better than the other things we're doing, after all these years," says Stephen Badylak, who runs the McGowan Institute for Regenerative Medicine in

Pittsburgh. So Levin is trying something completely different. Instead of trying to micromanage the wildly complicated universe of molecular and chemical interactions involved in building an appendage out of individual cells, he thinks it's possible to instead turn on the bioelectric switches that shaped that mouse (and all its fingers) in the first place. He's banking on the idea that the ability to regrow anything you've lost to injury or illness is not written in your genes, but can be controlled by the electrical language the body uses to talk to itself about what shape it is. Figure out that code, and you can just get nature to build you a new one. The first hints of these electric switches' existence go back nearly a century, long before we knew what to do with them.

The spark of life

Had he tried to undertake his experiments today, Harold Saxton Burr would have been frog-marched straight to HR. But in the 1930s, it was still conceivable for the director of a Yale biology lab to ask the women who worked for him to measure their voltages every day and chart them against their menstrual cycles.

Burr spent his entire career at Yale University's medical school, where his prolific publishing record bracketed the mid-twentieth century. His life's mission was to understand whether all biological systems exhibit electrical properties, and if so, why. To catalog the full extent of biological electrical activity, he spent thirty years wiring up everything from bacteria to trees to women, measuring and mapping the subtle forces they emitted. When Burr started this project, both the EMG (electromyograph—for muscles) and the ECG (electrocardiograph—for the heart) were already in wide use. But he wasn't interested in these loud and obvious rhythms.

Hidden among all this noise, Burr had identified a different signal—a faint electrical signature that never waxed, never waned, only persisted. He wanted to know more. To pin down this signal, he first had to spend three years devising a millivoltmeter so sensitive it made the heartbeat Augustus Waller had managed to detect with his buckets seem like eavesdropping on a gunshot.[4]

For his initial investigations, he asked the men in his lab to submit to voltage readings. Two electrodes were stuck into cups filled with an electrolyte solution, into which the men would dip their respective index fingers to get a sense of the difference between the voltage at the two fingers—a bit like what Ken Robinson did to demonstrate the wound current, but no one had to slice anything open. Still, Burr's extremely sensitive voltmeter registered a difference. "Immediately it was clear that there was a voltage gradient between the two fingers," he wrote.[5] This steady DC electric field, he realized, was evidence that the men all carried their own personal electrical polarization—one side of our body negative, the other positive. He called it "the electrodynamic field," or L-field. It was the first evidence that we are a human battery.

To be sure this signal was real, Burr and his colleagues repeated the experiment ten times (and with variations to rule out misinterpretations). When they were satisfied, they began their study in earnest. The men were instructed to take these measurements every day of every week of every month. Examining the results, Burr found that he could plot different men's field strengths along a spectrum. Some men consistently exhibited robust voltage gradients as high as 10 millivolts while others barely cleared two, but an individual man's field never varied much day to day.

That was when Burr began to wonder about the women in his lab. Might their signatures be more variable? He asked them to join the experiment.[6] And sure enough: "we found to our astonishment that for 24 hours every month, there was a large voltage increase." This coincided—after an "examination of the personal records of the females"—roughly with the approximate midpoint of the menstrual cycle, suggesting at once that the rise might be associated with ovulation.

Even in the 1930s, you couldn't have taken this experiment much further with human women, so Burr tested his hypothesis in a rabbit. A rabbit's ovulation is predictable: stimulate its cervix, and nine hours later an egg drops. They did a rather gruesome experiment to be able to read the voltage of the rabbit's ovary while simultaneously and directly observing the actual event of ovulation, opening its abdomen and extruding the fallopian tube.[7] "To our delight the moment of rupture of the follicle in the release of the egg was accompanied by a sharp change in the voltage gradient on the electrical recorder," Burr wrote. "The experiment was done enough times that it was perfectly clear that there could be no question that the electrical change was associated with the event of ovulation."[8]

Replicating this exact experiment on a living, breathing woman would have been out of the question. However, Burr was able to find a very close proxy: a young woman who was about to undergo an investigative surgery. She agreed to let them conduct their study, and so, during the fifty-six hours she spent waiting for her operation, they continuously measured her with their recording galvanometer. Burr placed one electrode outside onto her central abdominal wall, and the other inside, against the wall of her vaginal canal near the cervix, and watched for changes in the voltage between the two. When the recordings showed the same spike in voltage gradient that Burr had observed in the rabbit, the patient was sent immediately to the operating room for the laparotomy. The ovary was removed as planned, and close examination revealed a recently ruptured follicle—the sign of ovulation.

To Burr, this was clear confirmation that his findings in the rabbit translated to women.[9] He did a few more studies in this vein[10] and soon caught the eye of *Time* magazine, which in 1937 reported on "an electrical gadget whose invention may bring Dr. Burr a Nobel Prize."[11] The reporter described the gadget in loving detail: "In a box small enough to be carried around are four different kinds of electric batteries, a delicate galvanometer, two radio vacuum tubes, eleven resistors, one grid leak, and four switches."[12] Burr offered to share the wiring diagram with anyone thinking of building one of his devices for personal use, but cautioned the reporter that it should be assembled only by "an experienced mechanic who is thoroughly familiar with radio set construction." But it might be worth your while, as this complicated device could do something no one else had yet managed: tell you when a woman's ovary was about to produce an egg. *Time* dutifully explained that this would be a boon for people trying to start a family, but did not end the article without the delicate acknowledgment that "such foreknowledge might guide a woman's conduct in case she did *not* want to have a baby." Today, we can be a bit more direct: it could provide birth control.

Meanwhile, Burr's findings were confirmed in other animals by several scientists, including the Cornell-trained animal behaviorist Margaret Altmann, who found the same bioelectric correlates in sows and hens as they went into heat.[13] All this fuss finally drew the attention of John Rock, a prominent obstetrician and fertility specialist who ran the gynecological hospital at Harvard.

Rock got involved because Burr's hypothesis courted some controversy. At the time, it was presumed that all women ovulated like little clockwork dollies, dead center in the middle of their menstrual cycles—fourteen days before onset of menses, tick tock, out pops an egg. The data was not grounded in particularly solid science; it had come from epidemiological studies of veterans returning home after the First World War, and how fast their wives subsequently "fell pregnant." These observations had quickly graduated to the status of generalizable scientific knowledge.

Burr's findings indicated that while this "mid-cycle ovulation" rule might have been a good rule of thumb, individual women's monthly timings could vary, some quite widely. Indeed, his data suggested that some women ovulate more than once a month, that others had wildly variable fertile windows (never the same window in consecutive months), and that consequently it was very hard to get pregnant solely on the assumption that your fertile window was centered on fourteen days. On the other hand, it was a great way to get pregnant when you didn't want to.

Rock was a Catholic fertility specialist who had pioneered early sperm freezing and in vitro fertilization techniques. Quite out of step with the church, he was strongly in favor of women being in control of their own reproductive destinies, and would later go on to play a crucial role in the development of the first birth control pill, (vainly) lobbying the Pope for its acceptance.[14] But in the late 1930s, the only birth control even conditionally regarded as moral by the Catholic Church was the rhythm method, in which women would track their past periods to predict the time of the month they were least fertile (it required an ironclad faith that past performance was a guarantee of future results). Rock was in charge of a clinic where he taught clients how to use it.

For the rhythm method to be trustworthy, however, the average woman would need to ovulate at regular intervals; if a woman who actually ovulates on the twenty-first day of her cycle is only abstaining from sex in the middle of her cycle, you might accidentally end up with more little Catholics running around. When Rock saw Burr's experiments, he quickly set up a number of measurements in his hospital and completed experiments on ten more women to confirm Burr's findings.

Initial findings seemed promising, but Rock changed his mind within a year. After noting a number of discrepancies in the form of voltage deviations, he abandoned his investigation. Rock concluded that Burr's work had been misguided: there was no way ovulation could be taking place at these random times so far away from the center of the menstrual cycle. In his final publication on the matter, Rock dismissed Burr's findings about electrical signals and returned to the view that the deviations were anomalies from an otherwise reliable norm.[15]

Despite Rock's confidence in his own insights into women's reproductive machinery, today we know that Burr was right—the rhythm method is bunk. And later it emerged that some electrical changes are indeed well correlated with fertility. Chloride ion concentrations, for example, spike right before you ovulate.[16] That is so evident, particularly in the cervical mucus and saliva, that it has become the basis of an ovulation test which was developed specifically to check for concentrations of these ions. Examine these fluids under a microscope and you can literally watch the chloride crystal deposits flowering into crystalline patterns that resemble ferns.[17] These are bona fide indicators of fertility. (Anecdotally, when a personal friend of Burr's used his electrometric method after struggling with infertility, she was able to conceive. Burr wrote it up as a case study.[18])

Burr's early experiments run rather far afoul of modern workplace norms, but he was prescient: everything he theorized about bioelectricity in the body has been validated in the fifty years since he said it.

The electricity of development

Beyond the small flurry of replications undertaken in the 1930s and 1940s, no one has repeated Burr's studies of ovulation voltage. So we can't say with certainty exactly what signal he was detecting. But we do know, from nearly a century of other experiments undertaken since, that both eggs and sperm are electrogenic—living cells that produce electrical activity. A mind-boggling amount of it.

As both Burr and Lionel Jaffe would have told you, human eggs are far more difficult to study in their natural environment than the eggs of seaweed and frogs, which conveniently go through all their reproductive stages outside a womb. This is why so many studies of animal development have been done in frogs, and so few in humans.

While they're still sleeping in their follicle or testis, young eggs (oocytes) and young sperm (spermatids) don't emit strong signals. But as they mature, eggs of all species ramp up their electrical activity.[19] Just before an egg gets ready to drop off the mothership, it starts broadcasting energetically, almost like someone turned on an electrical switch. (The strength of this signal has been used to determine which eggs are best to use for IVF.[20]) Elisabetta Tosti, a biologist at Stazione Zoologica Anton Dohrn in Naples, found that this "on" signal is carried by a change in the amount and kinds of ions that flow through the egg's membrane, causing the egg to become hyper-polarized.

Sperm have a similar electrical on-switch that prepares them to meet the egg. In the 1980s, studies of sea urchin sperm found them to be teeming with potassium and chloride channels and the other usual suspects you find in neurons—and just as in neurons, blocking those channels prevented the sperm reaching their goal. For example, one of the most important electrical currents in human sperm is calcium, which confers an extra turbo boost to help it through the hostile terrain of the reproductive canal.[21] Take out the calcium channel and the sperm wriggles ineptly and goes nowhere. (This mechanism has been explored as a potential avenue for male birth control.)

Once it actually gets to the egg, you might think the sperm has one job, but it actually has two. We all learn in school about how it transports the male genome into the egg. For that to even have a chance of happening, however, the sperm first needs to hit yet another electrical on-switch on the egg's membrane. This one is known as "activation" and, genome or no genome, it's crucial for any further development to take place. It differs from the maturation on-switch in the same way switching on your bedside light is different from igniting the first stage of a spacecraft. The first touch of sperm to egg triggers an immense calcium current that smashes across the egg. Now no other sperm can get in, making it harder for silver-medalist sperm to get over the finish line.

This process is so consistent that when researchers zap a calcium current into an egg *without* sperm present, the egg gets excited and starts turning into an embryo anyway. That's right—virgin birth! By artificially mimicking the (normally) sperm-induced calcium wave, it jump-starts an egg into dividing without the sperm or its genome.[22] Ethical considerations prevent us finding out just how far this bootstrapping of the reproductive process would take a human embryo, but in rabbit eggs it got the embryo about a third of the way through development. (Fun fact: while not parthenogenesis, the special sauce in cloning Dolly the sheep, the world's first cloned mammal, was an electrical zap that activated the process.[23])

The point is this: through all the stages of conception, from egg to fertilization, ion channels and the currents they generate play a fundamental role in the spark of life. But none of this holds a candle to their importance in influencing the shape we eventually take.

Assembly instructions for one (1) human

A Lego set generally comes with a detailed, step-by-step instruction manual that leaves little doubt as to how you'll snap together each successive Lego piece. It also should give you a good overview of where any particular piece fits into the larger blueprint of the final structure you're assembling.

Making an embryo is much like making a Lego castle: in the same way that a castle needs turrets and gargoyles and a moat, you need two legs and two eyes and a heart. Except unlike the Lego Camelot, you don't come with a picture on the box of what you're meant to look like in the end, much less an instruction manual—and you're not going to be the one to assemble the structure. Instead, you'll sit back and wait for the Lego pieces to organize themselves. Our cells, our little Lego pieces, assembled themselves. What's even more astonishing is that when they got it right, all those cells got it right in broadly the same ways: we all managed to come out with the characteristic shape and proportions appropriate to our species (we can all spot a regulation-issue chicken, frog, mouse, or human shape).

So how did all our initial progenitor cells know how to organize into us—forming our eyeballs, legs, fingers, and all that in the right place and the right order? Who gave them the blueprints to check that all those fingers or fins or beaks weren't too giant, or too tiny, or of wildly different lengths? Most important—how did they know when to stop?

Well, you might be thinking, that's what DNA is for. Not true. You can search all the As, Ts, Cs, and Gs in your genome, turn 'em upside down and shake the spare change out of their pockets—you won't find the instructions for anatomy. You'll find plenty of specs—code that tells you about the color of hair a baby will have, its skin, its eyes. You'll find nothing about how *many* eyes. There's no gene for two eyeballs. There's no gene for "the eyeballs need to go on the front of the head." There's also no gene for "two arms and two legs, kind of this far apart." It's simply not possible to recover the shape of an organism solely from reading a printout of its genome.

So if not genes, what does control your shape?

The question had begun to assemble itself in Michael Levin's head when he was still a child, wondering how an entire person could be assembled from an egg. Later, puzzling over the old studies done by Lionel Jaffe and Harold Saxton Burr, he began to suspect that the ion currents Jaffe had found eddying around seaweed, and the fields Burr had measured coming off, well, everything else, might play a crucial and early role in determining a creature's anatomy. But where to start with a question that big?

As it happened, Levin needed a topic for his PhD thesis at Harvard Medical School, and in the early 1990s, there was still one aspect of how human beings are shaped in the womb that remained a total mystery: how embryos tell left from right. There had been theories, but never a smoking gun. For a grad student, this was tempting, low-hanging fruit. So Levin began to investigate how all those cells, without having brains, nonetheless seemed to know their left from their right. And make no mistake—their ability to distinguish left from right during development is critical to our survival. From the outside we may pull off the illusion of symmetry—two eyes, two ears, two arms, and two legs, same on one side as we are on the other. But inside, it's a different story. You probably know that your heart and stomach lean left, and that your right side houses your liver and appendix. For about 1 in 20,000 people, this whole image is flipped.[24] And that's absolutely fine! They don't usually have health issues (apart from overzealous researchers poking and prodding them to understand their condition, known as situs inversus).[25] However, when just some of the bits are flipped, you have a problem. Jumbling up the body's precise internal asymmetry, particularly when it affects the fastidious plumbing of the heart, is the origin of many congenital heart defects and other life-threatening syndromes.

Understanding what caused any of this—the correct pattern, the flipped pattern, the jumbled pattern—was a long-standing, perennially interesting unsolved mystery. Why does the heart go on the left and not the right? And how does the body know to develop this way? No one had been able to finger a specific molecular component, so there was no hint of a genetic cause. Genes couldn't be the whole story, anyway. After all, genetic information is not spatial. The genome can't tell left from right. It seemed to Levin, from studying the old ion current papers, that electricity was somehow fundamental to establishing the polarity of a cell. But how?

Jaffe was far from the only one who had investigated these questions.[26] Decades of work cataloged every ion that shuttled in and out of developing embryos of every species and identified the ion channels that send them through the zygote and the blastomeres it calves off into as it begins to diversify into the developing embryo. A funny thing happens to the ions and ion channels in the cell during this transition: they all mysteriously change. Some pop into existence, others disappear then reappear, and their currents wax and wane with those disappearing and reappearing acts.

Another clue to the functional importance of these weird ion events was what happened when you interfered with them. Disrupting even seemingly minor sodium currents resulted in a "rosette," which is an abnormal embryo that "appears to have lost its spatial orientation," noted the Italian biologist Elisabetta Tosti. She concluded that the currents during and after fertilization are crucial for correct embryo development.[27] Messing with potassium currents also led to developmental defects, further evidence that ion movements are crucial for an embryo. But no one had been able to assemble this jumble of interesting pieces into a coherent whole.

By the time the twenty-first century got underway, Levin could ask these questions in his own lab at the Forsyth Institute, at Harvard. How was electricity setting the polarity of a cell? He and Ken Robinson discovered a proton pump, another variety of the bouncers we met in Chapter 3. Protons are hydrogen ions. This bouncer specialized in making sure hydrogen and potassium were kept in strict ratios. On an unfertilized frog egg, proton pumps are speckled evenly around the whole surface.

But when Levin and Robinson checked on these pumps after fertilization, they found something strange: all the channels started drifting over to one side of the egg, where they smashed themselves into a tight little clique. No one had ever seen anything like this. When the pumps gathered on one side of the egg, it meant hydrogen ions were only able to get in or out of the cell on that one side. This created a voltage, and it happened really soon after fertilization, when the frog embryo consisted of only four cells. Could this be the answer they were looking for?

When scientists think they have found a causative agent like this, their next step is to try to figure out an experiment that can disprove their idea. Levin and Robinson decided to see what would happen if they prevented the proton pumps from drifting out of perfect symmetry after fertilization. To do that, they added extra proton pumps or potassium channels to the developing embryo to even out their distribution, mimicking the smooth distribution on an unfertilized egg. If the researchers were right, this evenness would wreak havoc on the embryo's ability to discern left from right. And they were correct. The embryos with the extra proton pumps were all messed up, as likely to have their hearts on the right as the left. The proton pump was clearly essential to kicking off the difference between left and right sides.

But it was also changing the membrane voltage. That was weird. As we saw in Chapter 3, a change in membrane potential is how nerves send action potentials. But why was a brand-new embryo changing membrane potentials? What possible use could that be, when it hadn't even developed nerves yet? Levin wondered if this voltage was part of the system an embryo used to tell its constituent cells to become different kinds of tissue. This idea is articulated by the biologist Mina Bissell: if all our cells have exactly the same genes in them, then why do some of them do one thing when others do other things? How do some of them become bone cells and others skin, or nerves?

The ghost frog

In 2003, Dany Spencer Adams was a restless assistant professor of biology on the tenure track at Smith College in Massachusetts. Having trained in the biomechanics of developmental biology, she had begun to find her job unfulfilling. After some sleepless nights, she decided to abandon the prospect of tenure, and take her chances doing something more interesting.

She saw a job ad for a post-doctoral position looking at left–right asymmetry. It wouldn't be a standard career path, but Adams was intrigued enough to drive to Boston to find out more. Within an hour, Levin offered her the job, and she knew she was going to take it.

Adams started with the proton pump that Levin and Robinson had found. Step one was to turn their finding into a tool that could take control of those ions to tune the membrane voltage of a cell. By tweaking voltages in frog embryos, she and Levin were able to create situs inversus, that mirror-image organ condition.

They began to notice that many of these tadpoles didn't just have inverted organ patterns—they also had very similar anomalies of the head and face. There was a clear pattern. This was dramatic evidence for Levin's hypothesis that these membrane voltages might be in charge of a whole lot more than internal asymmetry—what if they were in charge of the whole body?

To go any further, they would need to observe those changing membrane voltages in a way you could track with the naked eye. But what tool could allow you to see changing membrane voltages not just in space but over time?

Robinson suggested using an electro-sensitive dye that was able to turn differences in voltage into something that was plainly visible, in this case a gradient of brightness.[28] The extremes of the electrical potential were translated into degrees of luminance, with high voltages represented in bright white, low in black, and anything between in a balayage of gray. This dye could be infused into every single cell and track every one of them, even as they divided and proliferated. Using this trick, Adams would be able to watch every electrical step of embryo development.

Remember how I said that neurons rest at about 70 millivolts more negative inside than the outside? That's what the textbooks tell you because that's true of neurons and many other mature cells—but it's not true of embryonic stem cells (the little guys who proliferate during the first stages of development). Stem cells' resting voltage is much closer to zero. (That means the charge inside and outside their cell membrane is about the same, which is also the voltage of a nerve cell in its "panic at the disco" moment.) But where that zero moment is only momentary for a nerve, it is the stem cell's permanent identity.

Until, that is, it turns into something else. And that role is reflected in a cell's electrical potential.[29] You already know about the nerve cells' potential (-70). Skin cells have the same potential. But skeletal muscle has a higher potential, a firm and immovable -90. Fat cells are a relatively wobbly -50. What they all have in common is that they use their ion currents to keep their membrane voltage at the resting point that defines their cellular identity. A stem cell's low potential ensures that it can become any other cell. But once it has become a bone, nerve, or skin cell, it stays there. It gets set in its ways, a bit like us.

With electrosensitive dye, it was possible to watch all these electrical becomings unfold at the same time, in real time. Different regions of cells lit up at different times, forming patterns that faded into and out of existence on the embryo's surface. Many patches of embryonic cells would be near zero. But at any given point, some patch might drift up to -30. Another might hit -50. You could watch each region slowly glow up like Lilliputian cities coming online. It was beautiful to watch, but it didn't lend itself to any overarching theories.

Then, one autumn evening in 2009, after a day of watching these embryonic glimmers, Adams decided to leave her camera on record for the whole night. Her expectations were low; these little developing embryos would probably start wriggling around, leaving her with footage that was blurry and unusable. But what she found on her return the next morning was a "jaw dropper."[30]

On the otherwise featureless, smooth blob of a frog embryo, the hyperpolarized (negatively charged) areas twinkled brightly against darker areas of depolarized cells, as they had before. But then, as the froglet continued to develop, the random bright patterns playing across the dark surface suddenly cohered into a picture that looked an awful lot like a couple of eyes over a mouth. And then, sometime after those shimmers had faded, real physical features began to manifest in their place. Exactly where the electric glow had presaged eyes, soon there were two actual eyeballs. Precisely in the place where the pattern had projected the ghost of a mouth, development began on the real thing.

Soon all kinds of features developed exactly where she had seen their electrical premonitions. Not only could you match the voltage patch to the tissue, it perfectly predicted what kind of tissue would form, and its exact shape. It was stunningly clear: electrical signals appeared to encode the locations of anatomical features.[31]

The next question was pretty important—were these signals *necessary* for a normal head and face to form? Or were they just irrelevant indicator lights? To find out, Adams and Levin would need to prove that normal development was affected if you turned off the electricity. When they disrupted the ions that were responsible for the predictive patchwork quilt, that's exactly what happened: not only did that lead to changes in gene expression, but after removing the paint-by-numbers pattern indicators, the faces that emerged from the electrical chaos were deformed.[32]

So what exactly were they disrupting? And how was it possible that these brand-new, unformed cells were able to talk to each other about their voltages, or what parts to form? How were the membrane voltages spreading from cell to cell? Well, remember gap junctions? They start to form the moment the zygote has come together—that first new cell created by fusion of egg and sperm. Right away, they establish a cellular intranet quite unrelated to the nervous system, connecting cell to cell to cell.[33] Each new cell that cleaves off is already connected to the cells around it. Long before nerve cells develop synapses, our non-excitable embryonic cells have another, much faster, more electrical way of communicating.

Levin had long suspected that these gap junctions were involved in how an organism decides how to shape itself. In his earliest days of investigating left–right patterning, he had found that turning off gap junctions also messed up asymmetry. Later, he and another postdoc, Taisaku Nogi, found gap junctions were the culprit in the unparalleled regenerative superpowers of a weird little sea worm called a planarian. This small flat worm can regrow no matter how finely you chop it—and it only takes about a week to return to fully functional normalcy. Nogi and Levin realized that gap junctions could explain how re-patterning information could spread so quickly over thousands of cells.

In two different animals, then, the gap junctions seemed to enable long-distance messages without a nervous system. In some ways, they were *better* than a nervous system. When two cells are connected in this way, each cell has direct, privileged access to the other's internal informational universe. What one cell knows or experiences diffuses immediately through the connecting door for its neighbor to know or experience too. The effect is close to telepathy.

It was becoming clear how the whole thing worked: ion currents controlled the membrane voltage. The membrane voltage determined which tissue group a cell joined, which determined what kind of tissue it turned into. Cells changed their identities in line with cues they got from their neighbors, and the whole process was kicked off electrically.

This was when Levin first began to formulate his theory of the bioelectric code. The membrane voltage carried information, and the gap junctions formed the body-wide network—the electrical network that was not the nervous system—that sent that information around the body.

Levin started to think of the information as taking the form of a code. That code controls the complicated biological processes that formed you in the womb, by executing a controlled program of cell growth and death. The bioelectric code is the reason you retain that same shape throughout your entire life; it prunes your dividing cells so you keep being recognizably you. It wasn't the only important thing—biomechanics, biochemicals, and all the rest mattered too. But just as the neural code was mooted to govern behavior and perception, and the genetic code governs heritable traits, the bioelectric code was how the body told itself about its form.

But if all that was true, he would need to prove it. He would need to show that changing those cues could make the cells do something they normally didn't. It would have to be something really bonkers.

While messing with a particular potassium channel in a tadpole in 2007, Adams and Levin and their grad student Sherry Aw inadvertently altered its bioelectrical signaling, causing it to grow two identical, extra right arms alongside the original appendage.[34] This had been an accident, though—could they now do it on purpose? Aw hypothesized that "for every structure in the body there is a specific membrane voltage range" that drove the creation of that structure.[35] They tested that idea in 2011, tweaking the membrane voltage on a patch of tissue on a developing frog's gut to mimic the same hyperpolarized state Adams had seen before eyes formed on the ghost frog. It worked. An eye grew on the frog's stomach. They did it again on the tail. Another eye grew. "You can put eyes pretty much anywhere on a frog by changing the membrane voltage," says Adams. "It's like an X marks the spot."

If it was possible to grow new eyes anywhere on a frog, what could you do with a human?

Make like a salamander and regenerate

We used to think that only some animals could regenerate themselves: hydras, salamanders, crabs . . . nothing as interesting as a mammal. But in the twentieth century, the formal study of regeneration revealed just how widespread the phenomenon actually is in the animal kingdom.

In nature, there seems to be no theoretical limit to what you can cut off and expect to get back, if you find the right animal: hydras—tiny freshwater organisms—can be cut to absolute ribbons, and the little shred will rebuild itself again into a fully functioning animal. The same is true of that freshwater flatworm we met before, called the planarian.

In fact, this is how they reproduce—they tear themselves in half (you thought you had problems).[36] If you had this capability, someone could throw a segment of your finger into the sea and a week later, it would have grown into an extra you. You can see for yourself, in fact: chop a hydra in half and the tail end will sprout a new head and the head end will sprout a new tail.

Sea stars combine the abilities of hydras and planarians. In addition to being able to regenerate a new body from a severed arm, some species can regrow their entire central nervous system from scratch. They've been known to tear themselves in half on purpose to start a family,[37] and they've also been known to use their own severed leg to beat off their enemies.

Then there are salamanders, which can regenerate a remarkable number of their tissues and organs, including their limbs, tails, jaws, spinal cords, and hearts. A frilly red version called an axolotl can heal anything on its body without scarring, including its brain. Frogs can regenerate entire limbs and tails (and even eyes) when they are tadpoles, but they lose this ability after their metamorphosis into a frog.

Same goes for humans—at least until you exit the womb. To riff off a famous phrase often attributed to Abraham Lincoln: we can regenerate all our tissues some of the time, and some of our tissues all of the time, but we can't regenerate all our tissues all the time. Our regenerative ability follows a schedule that is strictly dependent on age and body part.

A zygote is the regenerative equivalent of a planarian. Someone could slice it in two and the two cells would continue developing into identical twins.[38] That ability falls off quickly, but a fetus has impressive regenerative ability even so. Most fetal injuries don't leave scars, an insight obtained in the late 1980s when fetal surgery became routine.[39] After birth, however, the superpower disappears fast, with one exception. Until between the ages of seven and eleven (for obvious reasons there hasn't been a lot of experimental evidence to pin this down exactly), if you lose the tip of your finger, you'll probably regenerate it in full.

This phenomenon is not extensively documented in the scientific literature—and not for the pinky-decapitating reasons you might think. Ai-Sun Tseng, a professor at the University of Las Vegas who leads a lab that specializes in regeneration, recalls describing her work to a class. One of her students "totally lit up. He was like, 'Yeah! Look at my fingers!' He grew up in the Philippines and at some point he'd had four of his fingers chopped off above the knuckle," she says. Because he was not yet eleven when it happened, they all grew back perfectly. But his age wasn't the only factor. His family had been too poor to afford a doctor, so they kept the wounds wrapped and wet and clean—and eventually all four fingers regenerated perfectly, nails and all. By the time Tseng inspected them decades later, they were indistinguishable from fingers that had never been maimed. At a conference a few years later, Tseng recounted the story to a group of colleagues, one of whom was a pediatric surgeon. He pointed out that, faced with a similar situation, most parents actually refuse to take advantage of this last vestige of regenerative ability. "They're way too scared of leaving an open wound," he told her. "They worry it'll get infected." So they ask the surgeon to suture together the surrounding skin, which protects the wound with fibrous scar tissue that forecloses any hope of the finger regenerating according to its potential. "Part of the reason we know about childhood regeneration at all is because of children in developing or poorer countries without healthcare," she recalls him telling her.

Our regeneration schedule is age-dependent—but it also depends on the body part. The liver renews about once every two months. Your intestinal lining sheds completely and renews again every seven days; what you eat next Saturday will be processed by a completely different set of cells than the ones working on today's breakfast.[40] A small population of stem cells in your lungs regularly undergo cell division. Even the lens of your eye regenerates. As you age, however, all these tissues lose their ability to rise from the dead, exemplified by the skin, whose outer layer renews every fourteen days while you're in your teens, but slows to twenty-eight or even forty-two days by late midlife. And of course, most of our tissue doesn't do it at all. Cut off a nose or a hand and they stay gone.

But why, when we obviously carry the genetic instructions for regenerating? Why can children regrow a fingertip but not a nose? Over the past couple of decades, the consensus across multiple disciplines has become that this latent ability actually lurks in every animal—and with it, our ability to regrow lost limbs or regenerate other organs. But how to unlock it? Once again, we turn to electricity.

Hacking the body map

Lionel Jaffe had found big differences between the electrical currents emitted by animals that regenerate limbs and the ones that just put a scar on a wound and call it a day.[41] In the early 2000s, Betty Sisken at the University of Kentucky painstakingly copied the exact qualities of the electrical fields that had been observed in regenerating animals and inscribed these onto the tissues of animals that don't regenerate. After amputation, a range of her animals—including amphibians, chick embryos, and rats—began to form limb buds. These had complex tissues like cartilage, vasculature, all the things you need for a functioning limb.[42] But alas, not an *actual* functional limb. Then Ai-Sun Tseng, who at the time was part of Levin's lab, manipulated the membrane voltages with ion-channel tweaks, and now we were cooking with gas.

She and Levin had been kicking around an idea—instead of micromanaging the process of regeneration, could it be possible to tweak the bioelectrics to kickstart the development processes that had built these appendages in the first place? Tseng started looking for ion channels that might be amenable to tweaking. She discovered one kind of sodium channel that was crucial to regeneration. Better yet, an ion-channel drug had already been developed that could act on these. Called Monensin, it was able to ferry extra sodium into a cell. Tseng had a hunch that flooding the cell with sodium—mimicking the electrical differences Jaffe had identified all those years ago—might restart regeneration in an animal that normally doesn't: a tadpole. Not only did it work, it worked shockingly quickly. Soaking for a single hour in the sodium channel drug bath drove eight days' worth of tail regeneration. When she told Levin about it, even he was skeptical. An hour seemed so short. But Tseng had been right. That little soak was enough to give the cells the idea, he says: "Grow back whatever belongs there."[43]

This was literally what he had envisioned being possible with the bioelectric code. Tseng had shown that all the persnickety chemical gradients, transcriptional networks, and force cues needed to orchestrate individual cells into complicated tissues could be harnessed with a comparatively simple set of electrical instructions. The genes were hardware, and they could be controlled by manipulating ion flows—the instructions from the software. Tseng and Levin soon published the seminal paper introducing their new idea: "Cracking the bioelectric code."[44]

Subsequent research has yielded multi-limbed frogs and other evidence of bioelectricity's role in regeneration. Among the most startling of these, it was possible to use bioelectric interventions to make planarians that had been chopped in half grow a second head instead of a tail. And as the press loves nothing more than a mutant, all the resulting media attention translated to money. First, DARPA came calling with enough money to build the little regenerative boxes that are now on the mice in Levin's lab. They've extended out to frogs, too, growing a new leg on an adult frog. The new leg wasn't perfect, but it worked—the frog used it to swim around, and after a few months it even regrew toes. In 2016, Microsoft billionaire Paul Allen added nearly $10 million to Levin's coffers.

The open question now: when will it jump to humans?

Electrifying regenerative medicine

Stephen Badylak heads up one of the largest projects on regeneration yet undertaken. It involves fifteen different investigators from multiple disciplines at eight separate institutions, and is funded by the US Army (which, if you want to be cynical about it, is uniquely incentivized to help heal the soldiers it has thrown into the geopolitical shredder). The goal is systems that can comprehensively understand the physiologic state of injuries, at every level from gene expression to mechanical properties, and then alter those states so the healing mimics development rather than default scar tissue formation. "*Star Wars*–type stuff," Badylak says. He is convinced that bioelectricity will play a role.

Bioelectricity researchers are considered the weird kids at the regenerative medicine table. Their paradigm is not entirely in step with early twenty-first-century science, which is heavily focused on genetics as the primary driver of human physiology. Every newspaper article about Levin's work includes a quote from some nonplussed geneticist mouthing a variation of "well, I guess we'll see." Most of the excitement is still focused on traditional avenues like tissue engineering and genetics, which are informing most of the human trials and work on lab-grown organs. Against this backdrop, work like Levin's can raise some eyebrows.

When Levin's team started publicizing their experiments a little over a decade ago, many biologists were openly hostile to the notion. Today, things are beginning to shift, as more traditional researchers start to delve into the specific relationship between bioelectric patterning and the genes it turns on or off. For example, Christiane Nüsslein-Volhard, who in 1995 won a Nobel for her work on genetic control of early embryonic development, is now among those investigating the electrical dimensions that seem to influence how zebrafish get their stripes.[45]

It should be said: regenerative medicine could certainly use the help. Organ transplants require an often lifelong regimen of immune-suppressing drugs to stop the body rejecting the new organ, which has its own health consequences. Metal replacement parts can become loose with time, engineered tissue scaffolds become inflamed, and artificial skin has no sweat glands or hair follicles.

In a perfect world, all these problems would have been solved by those famous stem cells. But despite the media hurrah, they have been a bit of a disappointment. The challenge has been how to stimulate them to become the cells you want them to be, and get them to go where they are needed, and keep them there in their new shape. Currently, most of the research on how to do that focuses on biochemical control. But we haven't had much luck with anything on the wish list: identifying, growing, inducing, or safely delivering stem cells to the appropriate target. In fact, it's rather unpredictable what will happen to stem cells once they get into your body.

This is why stem cells are regulated as an experimental drug, and the problem is highlighted by some fairly grisly anecdotes. One woman who had olfactory stem cells injected to heal her spine after a car accident ended up growing the precursor to a nose in her spine.[46] Another patient, who had stem cells injected in order to rejuvenate her face, ended up growing bones in her eyelids that were so big they clicked whenever she opened or closed her eyes ("a sharp sound, like a tiny castanet snapping shut").[47] After they started to interfere with her ability to open her eyes, she had an operation to remove the bones, though there is no guarantee more stem cells are not waiting in the wings with more castanets. Then there are the three women permanently blinded by a poorly controlled, poorly designed trial that harvested cells from their body fat to improve their vision.[48] Such examples are among the reasons stem cells for regeneration are banned on US soil, though of course they thrive in shady clinics, leading to regular warnings issued by regulators and other authorities about the "Wild West" of private therapies.[49]

But bioelectric medicine could offer a way out of this cul-de-sac. Preliminary work by Sarah Sundelacruz—a former protégé of Levin who was immediately snapped up into private industry—suggests you could tweak stem cells' bioelectrical parameters to influence their eventual identity. More recently, Sundelacruz showed that you could even analyze stem cells' bioelectric profiles to determine if they were any good at keeping their shape, or if they would revert into a kind of cell you didn't want—perhaps staving off the fate of the lady with the clicking eyes. This approach can even be used to guide stem cells into the specific physical locations where they are needed: Min Zhao's team have used electrical stimulation to guide stem cells to grow into replacement neurons in brain-damaged areas, which has been nearly impossible before.[50]

But what happens when the bioelectric signals that shape cellular identity go wrong? The consequences can be deadly.

CHAPTER 8

At the end:
The electricity that breaks you back down

The wound that does not heal

In the late 1940s, the zoologist Sylvan Meryl Rose toiled in his lab at Smith College, creating cancer chimeras. He would cultivate rapidly growing kidney tumors in frogs, excise them from their hosts, and then carefully graft the growths onto the legs of salamanders, tucking them just under the skin. (As we learned in the prior chapter, except for some brief periods during development, frogs can't regenerate, but salamanders can regrow whole limbs.) After the tumor transplant, the poor salamanders usually died of the resultant malignancies, with one exception: if Rose cut off the leg onto which he had grafted the tumor, precisely bisecting the implanted tumor, the animal regrew its leg, every time. The regenerating limb bud conscripted what was left of the tumor, transforming its cancerous cells into the normal cellular building blocks of biological tissue.[1] The regenerating leg essentially absorbed the cancer.

His experiment was among the first to identify the strange link between regeneration and cancer, but it was not the last.[2]

Among the strangest of these is the discovery of the naked mole rat's trifecta of superpowers: not only does this rodent rarely get cancer, it seems to heal without scarring[3]—and it defies known biological laws of aging.[4] These animals can live up to thirty years in captivity (a standard non-naked, non-mole rat clocks in at about a year). Naked mole rats had been known for a long time to be almost totally impervious to tumors, but in 2018 it emerged that they don't die of old age the way other mammals do. Emerging evidence also suggests that they can heal better than other mammals.

This odd story is one of many strange links between healing wounds, regeneration, and cancer. We have known since before Jaffe and Borgens that differences in bioelectric signaling are a critical component of both wound healing and limb regeneration—but instead of creating more of what we needed, what if it could also create more of what we *don't*? It would be a long time, however, before the proper study of the complicated relationship between electricity and cancer could be undertaken. The first scientists who tried to investigate the electricity of cancer faced an uphill journey, thanks to the long parade of Victorian con artists who had poisoned the well with their electrical cancer cures.

An indicator light for cancer

Around the time Sylvan Rose was chopping off salamander legs, Harold Saxton Burr and his colleagues got a visit from Louis Langman, an obstetrician at Bellevue Hospital in Manhattan. Langman hoped Burr's electrical ovulation-detecting technique might help bolster his success rates at artificial insemination, as one has to be ovulating for the procedure to work.[5] Burr—having just emerged from his bruising fight with Catholic doctor John Rock over the electrical signals in ovulation—was happy to help and instructed Langman on proper use of the device. It turned out well; electrometric measures improved the rate at which Langman was able to help women conceive. But it soon became clear that this wasn't the sole reason he had approached Burr. What he really wanted to know was if this technique might also help him identify cancers in his clients' reproductive systems.

Burr was in. He enthusiastically sent one of his contraptions back with Langman to his wards, where, in an initial group of 100 women, he strapped one electrode to the lower abdomen above the pubis, and the other either on or alongside the cervix.[6] Women whose troubles turned out to be caused by ovarian cysts or other non-cancerous medical issues almost always had a positive reading. Women with malignant tumors, however, showed an electrical "marked negativity" of the cervical region every time.[7] Langman confirmed their diagnosis with a pathological examination. Cancerous tissues, it appeared, emitted an unmistakable electrical signature.

Langman repeated the technique in about a thousand women to see whether his results stood up. They did: 102 of his patients exhibited the characteristic voltage reversals. When Langman operated on them, he confirmed that 95 of the 102 had cancer.[8] Even more remarkably, often the masses had not even progressed to the point where the symptoms would have driven them to visit the doctor, never mind obtain a correct diagnosis. After removing these cancers, the electrical polarity shown on the electrometer would normally flip back to a "healthy" positive indicator—but it did not always. When it stayed negative, Burr and Langman suspected that this indicated that they either hadn't got it all, or the cells had metastasized. Somewhere in the body, a cancerous mass was still sending its nefarious signals.

What struck them as especially strange was that the electrode inside the genital tract did not have to be placed directly on, or even particularly near to, the malignant tissue for the anomaly to be detectable. It was like a distress signal was being sent over distances through the body's healthy tissue.

It's hard to evaluate any of these experiments nearly eighty years after the fact. But to all appearances, it sure seems like a potentially reliable, non-surgical way to detect malignancy was discovered in the 1940s—and then got memory-holed. Langman and Burr were happy to acknowledge that "the method employed in this study is obviously an adjunct to other diagnostic procedures, and in no sense should be considered as a substitute for them."[9] However, it was something—and they wrote rather plaintively that they hoped others would refine this fledgling technique to aid early diagnosis. In his memoir, published twenty-five years later, Burr noted with evident disappointment that no one ever followed up their literature or did any replications.

In hindsight, it's quite easy to see why. No one had any idea what could possibly account for a voltage difference in cancerous tissue. Langman and Burr's findings were poorly understood, and like most bioelectrical phenomena outside neuroscience, were ignored. And then, of course, four short years later, studies of electrical signals in biology became moot with James Watson and Francis Crick's announcement that they had discovered the double helix structure of DNA. Oncology began to reorganize itself around genes. Not long after DNA was determined to be the sole arbiter of inheritance, it became canon that anything that damaged DNA and caused mutations in it may also cause cancer. In the 1970s and 1980s, a vigorous search for the abnormal genes ensued.[10] It was not a good time to go against the grain of science.

"A story that sounds almost like science fiction"

In the 1940s, while Langman and Burr had been investigating the electrometric cancer diagnostic techniques, Björn Nordenström was furrowing his eyebrows, puzzling over the subtle anomalies he kept finding in the X-rays of his lung and breast cancer patients. As a diagnostic radiologist at the Karolinska Institute in Stockholm, he had used X-ray imaging to inspect the blood vessels inside lung cancer tissues. It was during these examinations that he started to wonder about the persistent and puzzling irregularities that kept showing up in his images.[11]

The images in the X-rays looked like spiky flares around the tumors and lesions.[12] Colleagues dismissed these as artifacts of the imaging method, but that explanation didn't satisfy Nordenström. By 1983, he had devised a theory. Like Burr and Langman, Nordenström had found mysterious electrical differences between normal tissue and tumors, and concluded that these were the result of differences in the way ions flowed around them, thereby the source of the flares he had found, which he named "corona structures." He believed that both coronas and the flow of ions that caused them were part of a body-wide electrical circulatory system that existed alongside our traditional vasculature, like an additional blood stream. This system transported the ions in our "conductive media and cables" (including the blood) like little weather systems in coherent circuits around the body. Our electrical circulatory system was not only as complex as the circulation of the blood, but also similarly implicated in all the body's other physiological activities. But because it was invisible, until now we had missed it.

Controversially, instead of publishing this hypothesis as a series of small articles in highly ranked journals, which is the way scientific theories are typically disseminated, Nordenström decided to skip all that. In 1983, he self-published the whole thing in the form of a book—an oversized, 358-page colossus which he named *Biologically Closed Electric Circuits: Clinical, Experimental and Theoretical Evidence for an Additional Circulatory System*.[13] No publisher would touch it.

Some researchers would, though. The book's introduction sported not one but three forewords—and a preface—from four scientists willing to stake their reputation on this unusual idea. "I feel little need for lengthy considerations of its scientific merit," harrumphed Université d'Aix-Marseille biochemist Jacques Hauton with unsatirizable French *bof*-ery. "Its full importance is today impossible to appreciate," he continued, representing "no less than a major point in the evolution of our understanding of biologic science."[14] The other contributors were similarly dazzled.

It had only been seven years since the first ion channel had been observed ferrying sodium in and out of cellular nightclubs. In the eyes of science, that's a millisecond, so the idea of bioelectricity in cancer was inconceivable. "The theory sounds flawed," a National Cancer Institute deputy director, Gregory Curt, told the *Los Angeles Times* in 1986. "Based on what we know about cancer biology, there is no evidence that changing electrical fields have any impact on a tumor."[15]

Nordenström, however, had already started treating patients using the principles of his biologically closed electrical circuits to (he claimed) interrupt the electrical signals that promoted cancer. He would place a positively charged electrode needle into the tumor and another negatively charged one into healthy tissue, and apply 10 volts to the tissue for several hours. This was repeated until the tumor began to shrink.

Nordenström told the *Los Angeles Times* that the patients on whom he was experimenting "have been rejected by surgeons and other physicians as having cancer too advanced to be treated."[16] Between 1978 and 1981, he treated twenty such hopeless cases. Thirteen died despite his interventions. But Nordenström insisted that many of the tumors shrank and even disappeared. A short description of those first twenty cases appeared in 1984 in the *Journal of Bioelectricity*.[17] He insisted that he had been too busy to publish detailed accounts in mainstream journals—and, in a move that was sure to make him no friends, told the *Los Angeles Times* that what he was doing was too complex for many of his colleagues to comprehend. "People say it's controversial because it's another way to say they don't understand."

Here, surely, was the pseudoscience trifecta: a theory completely out of step with current scientific thinking; a standoffish refusal to publish in appropriate venues; and insistence on treatment being administered before the method has been properly validated. Nordenström was showing all the hallmarks of being a quack. And yet! Very smart researchers just couldn't agree. "It doesn't follow the usual medical logic but it fits in with a number of scientific facts in many disciplines," Morton Glickman told the *Los Angeles Times*.[18] It had taken Glickman, a radiology professor at Yale School of Medicine, a full year to inch his way through the migraine-inducing explanations of the biologically closed electrical circuits. He had come out the other side a believer. "My feeling is there is a very good chance that it will [turn out to be] true," he said.

While Western science held its nose, Nordenström's mentions of unseen forces traveling around the body piqued considerable interest in the People's Republic of China. In 1987, he was invited to Beijing to demonstrate his technique at the Ministry of Public Health.[19] There's not much information about that meeting, but afterward the ministry wasted no time in mounting an aggressive education campaign to get Nordenström's technique taught at hospitals. Between 1988 and 1993, forty-two courses were convened to teach Nordenström's methods, attended by 1,336 doctors from 969 hospitals.[20] By 1993, they had treated nearly 5,000 patients. By 2012, the technique had treated over 10,000 malignant and benign tumors.[21]

The media was skeptical, if not hopeful. On 21 October 1988, the highly regarded TV news program *20/20* ran a segment about a surprising new approach to cancer.[22] "Reporting on exciting medical breakthroughs is a role *20/20* has carried out for years," host Barbara Walters prefaced, before describing the vetting process the show goes through "to make sure we don't give credence to frauds." It was an unusual bit of deflection from an otherwise impeccably confident anchor. She then went on to introduce "a story that sounds almost like science fiction. It involves the theory that electricity plays a very major role in the human body. It could revolutionize medical science. It might even provide a new way of treating cancer."

So what do we make of this? As Walters's unease reveals, it can be really difficult in the moment to distinguish the quack from the revolutionary, but clarity tends to set in within a couple of decades. Not so with Nordenström. He disappeared. He may have moved to China to continue his work, and by one arcane account he died in 2006.[23] Several of the researchers who staked their careers on his claims have also passed away. Most people have forgotten him . . . but only most. Some of the researchers I interviewed have quietly squirreled away copies of his book, which are a rare find. They were kind enough to send me photocopies of certain sections, because like Glickman they believe—off the record—he will be vindicated.

Whatever you make of his theory—if you can understand it—there were some basics we didn't comprehend then which have since been put on solid foundations. One of these is that ion channels are present in all cells. Their activity determines the membrane voltage of the cell and the tissue, and thereby the behavior of these cells and tissues. Even cancer.

Cancer? There's an ion channel for that

Mustafa Djamgoz was in the middle of his third pint down at the pub when he got the overwhelming urge to put a cancer cell into a patch clamp. He had never heard of either Burr or Nordenström. He wasn't even a cancer researcher (yet)—at the time, he was a neurobiologist at Imperial College London. It was the early 1990s and he had met up with some of his old colleagues one evening after a conference to knock a couple back.

They were pondering cancer's electrical behavior and getting nowhere when Djamgoz, who had spent his life studying ion channels, had his eureka moment. "Suddenly this big penny dropped in my mind: 'My God, no one has looked at electrical signals in cancer cells!'" He asked his friends for some cells and got to work. Djamgoz didn't know it at the time, but he was about to embark on the most complicated and frustrating seven years of his career.

Good thing, then, that he was no stranger to complexity and frustration. Djamgoz grew up in Cyprus, whose Greek and Turkish residents have long been engaged in various territorial feuds. The island seethed under British colonial rule from 1878 to 1960, so when Djamgoz was born, every corner of his neighborhood was still festooned with characteristic British red telephone and post boxes. Throughout his childhood he dreamed of attending Imperial College, and as a teenager taught himself to build a radio transmitter from scratch. He would shock himself up to fifty times a day in the process—and not always by accident, as he was quickly developing a fascination with the way electricity is interpreted by human biology. This unusual child was soon lured from sunny Cyprus by a scholarship to study physics at a soggy boarding school in Kent, a springboard to Imperial. The university had an established reputation in visual psychophysics, the branch of vision research that probes the way animals transform physical stimuli like photons into our subjective sensory experience of the world, like the color blue. Djamgoz built an electrophysiology lab for his mentor from scratch, down to the amplifiers, and in return received his doctorate.

He spent the next two decades studying the electrophysiological response of the retina. "The retina is a beautiful model of the central nervous system," he says. "You peel it out of the eye, you put electrodes in, you flash lights and you can see all those individual cells responding." He still remembers the first time he stuck an electrode into a retinal cell and flashed a red light at it. The cell obliged immediately, flopping down into a state of flabby depolarization, its inner voltage becoming the same as its surrounding milieu, letting ions drift in and out of the cell as they please. Then he flashed it some blue, and the cell responded in the opposite direction, hyperpolarizing—re-establishing the big electrical difference between its inside and outside, which meant the ion movements were once again tightly controlled. "This one cell knew the color it was seeing," he marvels. "And you knew because you could see the potential on the oscilloscope just go up and down."

Djamgoz was doing these experiments in the mid-1990s, at the dawn of scientific investigation into adult synaptic plasticity—the idea that the brain's ability to change its connections doesn't end with childhood, but persists into later life.[24] Indeed, Djamgoz's work provided evidence for the idea, using the retina as his model to assemble proof that adult retinal cells can change their connections and adapt to different conditions. The work earned him a professorship in neurobiology, and that's where he likely would have spent the rest of his career if it hadn't been for that fateful night at the pub.

Now cancer seized his attention. One of his interlocutors from that evening gave him a batch of cells from prostate tumors in rats. Back in his lab, Djamgoz subjected these to the same electrophysiological prodding usually reserved for the retina. He found them teeming with electrical activity—but not the kind he was used to seeing in healthy cells.

It had long been known that as healthy cells turn cancerous, they de-differentiate: this means that they leave behind their prior identities as bone or skin or muscle cells, and return to a primordial state resembling that of a stem cell. But unlike stem cells, which often dutifully transform into new identities and go where they are needed, cancer cells refuse to "grow up." They simply drift, proliferating and consuming madly, never contributing to the society of healthy cells around them. This de-differentiation was reflected perfectly in the electrical activity Djamgoz observed in the cancer cells. The cancer cells had traded their strongly negative electrical identities (-70mV) for the permanently depolarized "zero" existence of a perpetual stem cell. (This observation of his was not unique, but in line with decades of previous observations.)

But that wasn't the only electrical artifact that grabbed his attention. They were doing something else too, something far more perplexing. These depolarized cancer cells were somehow . . . spiking. "These were bog standard action potentials," says Djamgoz—but what business did these cells have with action potentials? These cells came from gut or skin cells, not nerve cells. And yet, the aggressive cancer cells had somehow gained this ability to spike like a neuron during their transformation from healthy cells.

But the spikes they sent were not the dependable, decisive spikes of nervous signaling. They were much more chaotic than that, waving and flickering and exhibiting an incoherent pattern Djamgoz had previously seen only in epileptic episodes. What were these weird action potentials doing in cancer cells?

Djamgoz knew they were unmistakably the work of a voltage-gated sodium channel, the same family of sodium channels that lets nerves send action potentials. No one had ever investigated whether changes in these ion channels' behavior could be related to a cell's transformation into cancer. Could these aberrant spiking channels be the reason tumors became aggressive and metastasized? This was the question Djamgoz asked in his first paper. He and his colleagues submitted it to *Nature*, the premier science journal in the UK. It was rejected instantly by editors who dismissed the observations as an epiphenomenon. But Djamgoz and his co-authors eventually managed to present their work at an obscure urinary-tract conference. It was enough to get their paper published in a minor but respectable journal.[25] It was 1993, and Mustafa Djamgoz was done with neurobiology. The retina was out. Djamgoz only had eyes for cancer.

The next seven years were spent on what Djamgoz calls a charm offensive: publishing a barrage of incremental advances, climbing up the ladder of minor journals to arrive at middling journals, and talking to anyone who would listen about electrophysiology, bioelectricity, and basic physiology. An ever-increasing list of diseases were being accounted for by pathological mutations of the various ion channels—including cystic fibrosis, epilepsy, heart arrhythmias, and even gastrointestinal diseases—why should cancer be exempt? "Your electricity is what helps you get up and move around," he remembers shouting at his oncology associates. "Well, it helps cancer cells get up and move around too!" He continued to harangue his colleagues as he worked to build a careful foundation to understand the precise role these channels played in metastasis.

The broad consensus on cancer is that it results from abnormal expression of genes—or at least, a cell's initial flip from healthy to cancerous is usually chalked up to genetic defects and mutations. However, that is not what ends up killing you. It's widely accepted that most cancer deaths happen when the cells invade the rest of the body.[26] This incursion is facilitated by a repertoire of basic cell behaviors in which ion channels are known to be crucial: moving, multiplying, attaching, and more. It is not always possible to look at the genes in someone's prostate tumor and conclude from the DNA whether that tumor will sit there not bothering anyone, or whether it will start roaming around your body. But Djamgoz and his team were starting to wonder if maybe there was a clue in the action potentials. Could their spikiness correlate with the aggressiveness of the cancer? That would be a hugely valuable diagnostic tool.

By the turn of the century, people stopped dismissing these ideas. Other researchers had already been making the connection between ion channels and cancer, notably the Italian pathologist Annarosa Arcangeli, who had spent decades at the vanguard of linking cancer's electricity to specific genes.[27] At the University of Florence, she established the cancer-causing relevance of a gene called hERG, which is already familiar to many biologists in an electrical context: the ion channel it encodes plays a well-known role in coordinating your heartbeat by controlling its potassium current.[28] Arcangeli and Djamgoz were careful and talented scientists, and as more researchers began to join their investigation, overwhelming evidence accumulated that ion channels were key players in cancer progression.[29] Suddenly it wasn't just an interesting academic finding or even a new diagnostic: here was a promising avenue for new treatments.

Ion channel drugs were now a plausible way forward for cancer treatment. About 20 percent of drugs on the market target ion channels in some way, variously blocking them or prying them open.[30] If ion channels were turning out to be important in letting cancer proliferate, could blocking the right one help stop it? Could one of the existing ion channel drugs hold the key to stopping these aggressive cancers?

There was just one problem: the property Djamgoz identified as making cancer more aggressive was controlled by the same voltage-gated sodium channels in charge of the action potential. You couldn't block them. Sure, you might stop a person's cancer metastasizing, but you'd also stop their nervous system; bad news for their heart and their brain.

This is one of the hardest, most bedevilling problems in cancer treatment: finding some unique target that exists exclusively in a cancer cell but that will not also mess up a normal, healthy cell. "There's a huge history in cancer of people identifying some property of cancer cells," Mel Greaves told me. "But when you dig deeper, you often find those properties are not cancer-specific at all; it is only that cancer cells are exploiting a perfectly normal property." Greaves, at the Institute of Cancer Research in London, is cancer-research royalty—he received a knighthood in 2018 for his investigations into what triggers leukemia in children.[31] This is the guy journalists call when they want to know if something is legit in oncology.

But Djamgoz dug deeper. When he did, he found that the guilty cells were using a special type of ion channel that normally only existed in the cells of developing fetuses. There, they supercharged cell multiplication and the other processes needed to quickly form a whole human from nothing. By the time a baby is born, though, this turbo version should have been shut down, deleted, and replaced with the normal, "adult" version of the channel, the one that only does approved activities like sending action potentials.

Djamgoz's prostate cancer cells were teeming with these pre-birth ion channels, which he called an "embryonic splice variant." Something had woken them up again as the previously healthy tissue turned cancerous.

Now that Djamgoz knew how the aggressive splice variant differed from the normal and life-sustaining regular sodium channel, he had a target whose removal would not harm normal bodily operations. Over the next few years, he searched for the same variant in other metastatic cancers, trawling through biopsies from human cancer patients, and reliably found his splice variants (or their counterparts) in colon, skin, ovarian, and prostate malignancies.[32] This time, it did not take much convincing to get a grant from Cancer Research UK to work on an antibody to specifically inhibit these variants.

Mustafa Djamgoz and Annarosa Arcangeli are no longer struggling to get people to accept their ideas. More than two decades after Djamgoz's ion channel was dismissed as a coincidence, the field exploring ion channels and cancer has exploded.[33] Researchers around the world[34] are busy digging for hidden treasure in that big catalog of existing drugs.[35] What's more, sodium and potassium channels are no longer the only game in town. People are also looking at chloride and calcium. The picture that is emerging is of many different types of channels all working together in complicated synchrony, like an orchestra, as Djamgoz put it in a 2018 interview. The sodium channel "could be the lead violinist, but to be able to create the full symphony we must understand the other players as well."[36] For example, Arcangeli's hERG channel is now an object of great interest among pharmaceutical companies. At a roundtable of *Bioelectricity* editors in 2019, she predicted that novel therapies targeting ion channels would be a future cancer treatment.[37]

Djamgoz has his own company now, and they were starting to put together a human clinical trial when—as with so much science—the pandemic put everything on ice. Neither this nor the fact that he is not a clinical oncologist have stopped forlorn people from calling him at all hours of the day. "They are desperate," he says. People diagnosed with cancer need new options.

A new ally in the war on cancer

The efficacy of the most common treatments depends on the cancer being caught early, while it is still a tumor sheltering in place. Once the cancer has spread its tendrils elsewhere into the body, survival rates start to go down. Mel Greaves outlined his theory explaining why in the journal *BMC Biology* in 2018: successfully destroy a tumor with radiation or chemotherapy, and in theory, you've won. If no cells remain, for the moment, you are cancer-free. However, should even a single cell survive, then by definition, it is now immune to anything you threw at the tumor before. That cell is the mother of your future tumor, and as it proliferates, all its progeny will be equipped with that same resistance. (The same logic governs drug resistance.[38]) And there's evidence that this new batch of cells will be not only more tenacious, but more aggressive than the original tumor. "We are battling natural selection, one of the fundamental laws of the universe," Charles Swanton, an oncologist at the Francis Crick Institute in London, told *New Scientist*.[39]

To start working out the new battle plan, in 2013 Mel Greaves set up the Centre for the Study of Evolution in Cancer. He gave a talk at London's Science Media Centre, where he floated a new idea to address the resistance problem: for some advanced cancers—particularly in older patients—instead of hunting down every last cancer cell in the aim for a cure, perhaps we should approach it more like a chronic disease. "Most cancers hit people when they are past the age of sixty," he explained to me, recounting the talk he gave there. "If you treat cancer as a chronic disease, and keep it from becoming aggressive, you might get ten or twenty more good years." This would be a vast improvement over the mere months some treatments add to people's lives in pursuit of curing their late-stage cancer (not to mention the impoverishing expense, and the toxic medications that more often destroy their quality of life than their cancer). Not everyone was convinced. "I got a lot of hassle for that," he recalled in our conversation. An editor from *The Times* told him that it was the worst idea he had ever heard. The published responses were similarly unkind. "Let's stop trying to cure cancer, says cancer professor," sneered the *Daily Telegraph*.

But time has been on Greaves's side. A lot of scientists today agree: catch it early, but if you can't, "control is a much more realistic objective."

Genomics has revolutionized cancer treatments and has vastly enhanced our in-depth understanding of cancer. It has yielded powerful new diagnostic and therapeutic tools that have been incredibly effective in some cases, including a game-changing treatment for adult leukemia.

But there is a gulf between these successes and the contention that cancer is a disease of the genome. "Cancer is not purely a disease of the genome, just like evolution is not just about genes," Greaves says. A cell can change many of its attributes on the fly, in response to its environment, in a way that their genomes can't fully account for. "So saying it's all genomics is wrong," Greaves told me.

So the question is: if the electrome affects cancer, what can we do with that information?

AT THE END

Detecting the electrics

In the decades since Harold Saxton Burr and Louis Langman first suggested using cancer's electric properties to detect it, many research efforts have discovered that you can use bioelectric properties to distinguish cancer cells from their healthy counterparts, owing to the way they disrupt the flow of electrical currents through the body. This was an unfamiliar concept when Burr and Langman were doing their work, but is now widely known as bioimpedance.[40] You might recognize the word from those fancy scales they have at gyms and spas that measure body composition (though most people who use them are primarily interested in the precise ratio of body fat to muscle they reveal). These work on the principle that current can't travel through fat cells—fat having a higher "impedance"—but can travel through lean tissue like muscle. Cancer also has its own bioelectric signature.

When removing a cancerous tumor from any area of the body, the surgeon's goal in the operating theater is to leave nothing behind. But she is cutting blind, unable to see the difference between cancerous and healthy tissue. While imaging techniques and other technologies give a location and map of the mass, when it comes to the actual act of cutting tumor out of flesh, it's highly educated guesswork. To raise the odds that the whole mass is cleanly removed, the surgeon aims to carve out not just the tumor but also a generous rind of normal tissue around it, often several centimeters' worth.

After surgery, that carved-out lump of flesh is sent to a pathologist. The pathologist examines it—specifically, the rim of healthy flesh around the tumor, which is known as the surgical "margin"—to be sure that margin is clear of any cancer cells. The problem is that the results can take several days to come back, and if the analysis finds a positive margin—cancer cells are present in the rind—it will mean the patient needs a second or third surgery, and more treatments to augment them.[41]

Several new technologies, in various stages of clinical trials, aim to help surgeons get the whole tumor the first time. One promising candidate, ClearEdge, developed by a start-up in San Francisco, used bioimpedance for breast cancer margin detection. It integrated the technology into a device called a "margin probe." This is used by the surgeon while the patient is still under anesthesia, after the surgery, to measure the bioelectric properties of the area around the tumor that has just been removed. A "traffic light" bioimpedance map helps the surgeon see where she missed a spot: red for cancer, yellow for uncertain, green for clear. It was clinically evaluated in several hospitals in the UK. In 2016, surgeons at the University of Edinburgh Medical School and Western General Hospital, in Edinburgh, successfully used the device to identify cancer in areas around the excision and reported that it can reduce the need for repeat surgeries.[42] It compared favorably to existing, more time-consuming ways of testing for cancer.

So, where's ClearEdge? Why haven't you heard of it? Mike Dixon, one of the surgeons who trialed the device, told me that while the technology was easy to use and its results were pretty good, the follow-up studies never happened. "The company was reliant on venture funding," he said. "The technology sounds great," he says, but so have many other margin probes that their team has been involved with. Some proved too elaborate, others not accurate enough, and some just disappeared.

Dany Spencer Adams is working on a way to make an affordable, accurate version that she says anyone can use, based on the same kind of bioelectric dye that helped her visualize the ghostly frog face. It tattles on the peculiar electricity of cancer cells in a different way—by lighting them up according to their membrane voltage, so cancer cells appear to be a different color than healthy cells. They don't do this in a live, open patient though—they do it with the mass they've already removed, and a very fancy piece of blotting paper. After excising the tumor, a surgeon will press this special paper to the margin of the mass to transfer the cells, put the paper in the dye, photograph it, and upload the results to a computer program. Within ten minutes, you have a heat map of the entire surgical margin—a paint-by-numbers landscape that tells you where you missed a spot. If they did, they can go back in while the patient is still on the table.

That's the idea, anyway. After testing it in many cells in petri dishes, and watching the voltage dye cause the cancer cells to light up dramatically, testing has begun on live tissue with promising results. However, it's not available yet. Clinical trials are always expensive, and sometimes investor goals can work against new devices, when cashing out of your start-up is more important than making it into a surgeon's hands. We're therefore still a way off this new wave of bioelectric diagnostics making it into the operating theater, where it could make surgery for cancer much more effective and reduce recurrence, not to mention the trauma and infection risk that comes from more surgeries.

Further out, it may become possible to check your cancer's bioelectric properties to find out whether you even need surgery to take the tumor out at all. Remember, genetic defects may have caused the initial cancer, but whether it grows or goes on walkabout is up to your body's bioelectrics. Not all tumors are aggressive—some are slow, and may go away on their own. In an as yet unpublished study, Djamgoz and his colleagues have collected more evidence that their sodium channel could in and of itself be a diagnostic marker for a cancer's aggression levels.[43] When his channels whip up ion currents, survival rates drop, he told an ion channel modulation symposium in 2019. This could help people make difficult treatment decisions, for example when evaluating the need for radical and life-altering surgeries and other treatments. "We have never seen metastasis where the channel was not present," he told me. Djamgoz's sodium-channel findings are also opening up some really unexpected new options for how we might treat the cancer we find.

Communications blackout

To prevent seizures, some people with epilepsy take drugs that close the sodium channels that spark abnormal action potentials in nerves. This calms the electrically overactive brain action potentials, making them less likely to cascade. Such drugs don't just treat epilepsy symptoms; they have a wide range of uses, such as for heart arrhythmias and some types of antidepressants.[44]

A little over a decade ago, anecdotes at clinics and occasional reports to the FDA began to hint that people who took these sodium-channel-blocking drugs seemed to have a lower risk of getting some kinds of cancer, and were more likely to survive if they did get it.[45] According to follow-up reviews, those kinds of epilepsy drugs seemed to be associated with lower incidences of colorectal cancer, lung cancer, gastric cancer, and blood cancers.[46] (To be clear: these are early signs, not smoking guns. None of this is enough data for anyone to start taking anti-epileptic drugs if they're not needed!)

The very preliminary story of these sodium-channel-blocker drugs, however, happens to fit very neatly into Djamgoz's theory. Added to this mosaic, Djamgoz's research suggests a mechanism for the mystery of how the sodium channel blockers might be keeping cancer at bay. The erratic action potentials sent by his splice variant create a way for the tumor cells to establish contact with each other and the cells in the immediate vicinity. "They are communicating with each other," he says. Blocking them would block this communication.

These trials are all in their earliest days, but if they pan out, there's good news: the process to approve these drugs to treat cancer could be very short. Djamgoz, Huang, and Arcangeli are among many researchers repurposing the troves of existing ion-channel drugs to keep cancer cells from communicating and acting on their environment. One big draw of repurposing existing ion-channel drugs is that you don't have to start from scratch with drug development—which can take decades—and that can dramatically accelerate how soon you see them in the clinic.

If such drugs can remove a cancer's ability to metastasize, Djamgoz thinks you can turn the disease into a chronic and manageable condition—exactly in line with Greaves's position that cancer should be treated as a chronic illness. "We advocate 'living with cancer,' rather like we can live chronically with diabetes and the AIDS virus," Djamgoz said in his 2018 interview. "Living with cancer means suppressing metastasis, since this is the main cause of death in cancer patients."[47]

But ion-channel drugs could do even more than that. Some very early studies have raised the possibility that, just as Sylvan Rose's regenerating animals were able to hit the undo button on a growing tumor, messing with the right bioelectric parameters could help us do the same.

The society of cells

In the last few years, an unambiguous general consensus has emerged that the solution to the cancer problem likely resides in new theories of cancer. In 1999, Ana Soto and Carlos Sonnenschein at Tufts University School of Medicine suggested exactly such a novel paradigm: what if we started looking at cancer not as a breakdown of individual cells but as a breakdown of cellular society? When individual cells get together, they form tissue, and that tissue is a kind of society. Proliferation is a cell's default state, they argued. So cancer is not sparked by one rogue cell gone wrong so much as a failure of the local environment to keep the cell's "natural instincts" in check.

Cancer, in this view, becomes a disorder of organization in the human body rather than a defect of individual cells. It was a beguiling metaphor, especially as it squared so well with the way cancer cells stop contributing to the body and decide to live on their own radically individualistic terms. Nor was it as radical a supposition as it first appeared.

A spate of more recent work has begun to more closely examine the significance of non-genetic factors on cancer's spread: things like tensile forces and biomechanics in the microenvironment, and their contribution to a tumor's ability to expand and invade its surroundings. In 2013, researchers at Memorial Sloan Kettering Cancer Center in New York wrote that "many studies have shown that the microenvironment is capable of normalizing tumor cells," that re-education of the cells around the tumor, rather than trying to get rid of them, "may be an effective strategy for treating cancer."[48] In other words, the healthy cells around the tumor are just as important as the tumor in determining whether the thing can spread. It's not just the cells themselves but something in their environment (the society) that is falling down on the job of regulating their behavior.

In particular, the evidence has recently started to point to the importance of the bioelectric signals that cells use to process information. The same kinds of weak electrical fields that coaxed healthy cells to crawl across a petri dish have also convinced cells from brain, prostate, and lung tumors to make the trek.[49] Such fields, of course, also exist inside the body—they're the consequence of the currents swirling around the cytoplasm and the membrane voltage of all our cells.

To sum it all up, the interactions between cancer cells and the surrounding bioelectric fields are increasingly recognized as an overlooked but crucial aspect of how cells make decisions based on the state of their neighbors. In this framework, cancer can be viewed as a failure of communication—a fault in the field of information that coordinates individual cells' ability to be part of a normal living system.

So if that's the case, is it possible to re-establish the communication protocols? This is fairly unconventional thinking when it comes to cancer, and yet has a growing number of adherents.[50] But as we get a better view of the varied roles bioelectric signals play in cancer, new possibilities are emerging. The new suite of tools enabled by a focus on the bioelectricity of cancer could lead to earlier diagnosis, its transformation into a chronic disease—and maybe even a way to convince cancer cells to hit the "undo" button.

If you recall, the cell's membrane voltage is closely related to (and can determine) its identity, from stem to fat to bone.[51] Manipulating this voltage made many remarkable changes to an organism: like that eye made to grow on a frog's butt. Well, it turns out the same factor that could sculpt an eye on a frog's butt could also tamp down a cell's will to become cancerous.

If the body's "societal" control over its cells is mediated by the signal of membrane voltage, then a good way to test this bold theory would be to see if, simply by changing the electrical voltage of a cell, you could cause a healthy cell to turn cancerous, or convince a cancerous cell to return to a healthy state.

Those were precisely the experiments researchers undertook at Michael Levin's lab at Tufts University in 2012. If bioelectric signaling was an important part of how cells communicate to work on pattern and coherence, they reasoned, and cancer represented a break in this multicellular contract, then interfering with cells' ability to send bioelectric signals should lead to cancer. After Levin's doctoral student Maria Lobikin depolarized normal cells, the depolarized cells began to act malignant.[52] It was evidence that bioelectricity is the informational glue that holds big multicellular structures together. Membrane voltage was "an epigenetic initiator of widespread metastatic behavior in the absence of a centralized tumor," she and her co-authors wrote.

The next year, another member of Levin's team, Brook Chernet, went a step further: could you use membrane voltage alone to predict whether cells would become cancerous? They tested the hypothesis on frog embryos that had been laced with human cancer genes to cause them to form tumors. Using the same fluorescent voltage-reporter dye that Dany Spencer Adams had used to watch the electrical development of the frog face, they were able to observe depolarized membrane potential in the tumors. And just as Adams had been able to predict facial features, the electric signal change just by itself could predict which cells would turn cancerous.[53] This experiment, they wrote, not only implicated bioelectric signaling in tumor formation, but suggested new approaches for anticancer therapies. That's because when they repolarized (and strengthened) the low-voltage cancerous membrane, the cells stayed connected to the society and ignored their own mutated genes' efforts to turn them cancerous. In other words, Chernet and Levin reduced the number of tumors just by repolarizing the depolarized cancer cells.[54] Chalk up another win for the bioelectric code.

By 2016, Chernet could not only stop new tumors from forming—he was able to "reprogram" existing ones back into normal tissue in tadpoles. Their tumors were advanced: they had already spread and formed their own blood supply. But when Chernet used light-activated channels (a technique known as optogenetics) to modulate the cells' resting potential, they stopped acting like cancer. "You can turn on the light . . . and the tumor goes away," Adams, who was one of the co-authors of the paper, told Reuters.[55] Electrically reminding the cells of their role in the rest of the tissue, Levin told me, appeared to snap them out of their midlife crisis and help them re-enter the society of cells. Bioelectricity overrides genetics. As these and other experiments showed, the voltage changes were not merely a sign of cancer. They controlled it.[56]

This is all fascinating, but it, too, is an extremely long way from your doctor's consulting room. Like all the recent results on bioelectricity, those in cancer are still early. Tadpoles are really, really not like us. Furthermore, repeating some of the experiments has flagged inconsistencies.[57] There's a lot left to do.

Like regeneration, however, it offers a very tantalizing prize: a control switch for more complicated biological processes. "The electrical communication among cells is really important for tumor suppression," Levin says. What's more, this control switch might also be amenable to existing pharmacological interventions. Like Djamgoz and Arcangeli, Levin is also looking at ion-channel drugs.[58]

In a little under a century, bioelectric signals in cancer have gone from an overlooked curiosity to suspicious quackery to a promising way to improve cancer detection and treatment. The more recent investigations have illuminated why Burr and Langman were right: cancer has characteristic electrical signatures that can be used to detect it. Indeed, these signatures may be just the beginning.

Nordenström may have been onto something when he tried to disrupt tumors with electricity back in the 1940s. There's now a flourishing and rapidly advancing line of research into destroying tumors with nanosecond pulses of cold plasma that are more precise and powerful than anything he had access to in his day.[59] This newfound ability to harness room-temperature lightning for medical purposes is fast changing the way we treat tumors, says Jose Lopez, who directs the plasma physics program at the US National Science Foundation. This is yet another bioelectric intervention to watch in the next ten years.

Many devices and technologies are now being recruited to interface with the body's electricity for regeneration, wound healing, and cancer treatment, and they join ion-channel blockers as the new vanguard of medicines.

But that's now. There are things in the pipeline that look nothing like these tools. They won't be made of metal. They will interface with us on a much deeper level. And they will likely be made of the stuff we find in the natural world, which runs on the same electrical programming as we do.

PART 5

Bioelectricity in the Future

*"We have been promised a future of chrome,
but what if the future is fleshy?"*
Christina Agapakis

The bioelectric code is only one of several facets we are beginning to discover about the electrome. All of them suggest that successfully engaging with our natural electricity won't be about control and manipulation so much as it will require interacting with it on its own terms. Understanding the full breadth of the electrome will also require much more than just a command of ion channels or an understanding of the nervous system. It will require a huge interdisciplinary effort and a critical look at how the structure of today's science itself can limit scientific understanding. It will also require a rethink of the materials that we use to interact with our electrics. Maybe it will even lead to a new way to think about the drugs we take and their effects on the electrome. It will be, in other words, revolutionary.

CHAPTER 9

Swapping silicon for squids: Putting the bio into bioelectronics

Frogs have been through a lot over these past 200 years of electrophysiology, from Galvani's grotesque puppeteering to Matteucci's body horror power source. But no one could have anticipated the next role they would play in the quest to unite biology and electricity. In 2020, frogs became the raw material for a class of organism that had never before existed in the evolutionary history of the world.

Well, their cells did anyway. A few thousand of them were scraped off frog embryos and then reconstituted into groups of roughly 2,000. Under some cleverly programmed tutelage, the clumps began to cooperate, moving and acting of their own accord, becoming—in the verbiage of their creators—"xenobots," literally "frog" (from *Xenopus*) "robots." This wasn't the average person's idea of a robot, but neither were these frogs, not anymore. They did not have brains or nervous systems so their ability to move and decide was outside the traditional accounting for how animals do this. They didn't have mouths or stomachs so they couldn't eat. No reproductive organs meant they couldn't make more. Joshua Bongard, the University of Vermont roboticist who had helped create them, called them the only thing he could: "these are novel living machines."[1]

Wait—*roboticist*? Why is a roboticist creating frog cell bots?

Robotics is changing. Where they used to be thought of as hard-edged appliances that occasionally (as in *The Terminator*) take on biological forms, now the line is blurring between biology and robotics as we learn more about both. After all, a robot is a programmable device that can manage information—and a cell is turning out to be that too. The xenobots' creators speculate that the tiny organisms might someday deliver drugs into targeted areas of the body, scrape plaque out of arteries, or clean up plastic waste in oceans. But perhaps the most important thing they offer is a rare glimpse into the possible future of the materials we use for robotics, electronics—and implants.

For years, researchers have toiled to find new and better ways to interface with our nervous system, but have been thwarted by the mechanical, chemical, and electrical properties of existing devices and their fundamental mismatch with our brains. Compared to the signals they are tasked with manipulating, these metal devices are rigid and bulky. "It's like playing the piano with a mallet," Andrew Jackson complained when I went to his neural interfaces lab at Newcastle University to get a better understanding of the future of brain implants. (His phrasing echoes Kip Ludwig's on deep brain stimulation—two researchers using this phrase is an interesting coincidence; if a third uses it, I might cry conspiracy.)

For the past decade or more, the limitations of metal devices have motivated an enormous project to create squishier, stretchier, more biocompatible materials that let our bodies electrically communicate with the foreign bodies inserted into them. This trend extends from tissue engineering to robots, which are increasingly being augmented by, or made entirely out of, synthetic materials such as hydrogel, a squishy polymer that is popular for soft robotics.[2] In the future, such nano-gloop-bots will supposedly swim through our bodies, making adjustments to errant tissues.[3]

As we gain a better understanding of the electrical instructions of biology itself, a sizeable contingent of scientists is beginning to wonder if the ultimate biocompatible material isn't just . . . literal biology. That's why researchers are now studying the properties of sea creatures, frogs, and fungi for their programmability and biological compatibility.

The rise and fall of electroceuticals

Around ten years ago, an astonishing breakthrough was reported in *Wired* magazine, and from there quickly raced through the rest of the media. The neurosurgeon Kevin Tracey had used an electrical implant in a research subject's neck to zap the vagus nerve, an enormous treelike projection of nerve bundles whose branches extend to and from the brain through a vast amount of the body. Electrically stimulating it had dialed back the excruciating symptoms of the patient's rheumatoid arthritis, an immune disorder with which he had suffered for years.[4] It was an extraordinary story: before the treatment the subject had been so debilitated, he couldn't play with his children, but the electrical stimulation was so effective he could return to work, throw his kids around, and even resume his favorite game, ping pong. (This turned out to be too much of a good thing—he overdid it and ended up giving himself a ping pong injury.[5])

Rheumatoid arthritis was far from the only immune system problem this intervention promised to treat without drugs or side effects: asthma, diabetes, hypertension, and chronic pain were also promising targets. "I think this is the industry that will replace the drug industry," Tracey told the *New York Times* journalist Michael Behar.[6] Soon, science magazines and newspapers teemed with the catchy new portmanteau that described this merger of electricity and pharmaceutical: the era of "electroceuticals," it seemed, was upon us.

What so bewitched those of us in the science press, however, was not only the promise of transcending drugs. It was the elegance of the new mechanism: you didn't have to mess with drugs and side effects, you just flipped a switch and the body would do the rest. That is, he had recently discovered that the nervous system could control a lot more than our motor nerves. It might be able to control inflammation and the immune system. He believed the circuit he had identified was only the first of many ways we would find the vagus nerve entangled with our every organ and cavity, and thereby able to govern any number of their functions. The immune response had previously been considered beyond the reach of the nervous system's control architecture—we simply didn't realize nerves went there or did that. But now the list of ailments targeted for electroceutical intervention expanded to include chronic obstructive pulmonary disease (COPD), other heart conditions, and gastrointestinal diseases. They just needed the wiring diagram.

To find it, the global pharma giant GlaxoSmithKline set up a $1 million prize. The endgame, they explained to me in 2016, would consist of rice-sized electrical implants that would sit on specific control branches of the vagus nerve, monitoring messages as they flashed through on their way between brain and viscera: muting some, amplifying others, and generally recording the electrical activity within to catch problems and quickly fix them. It sounded like an NSA wiretap, but for your health. By then, Google's life sciences arm Verily had also become keen, and the two superpowers spun off a new supergroup, a venture they christened—remarkably—Galvani Biosciences. Early pilot studies confirmed the potential of the approach, with one group finding that the correct series of electrical impulses to the right nerve bundle could reverse diabetes in mice.

Surmounting the remaining technical hurdles "might take 10 years," Kris Famm, head of GSK's bioelectronics research and development unit, told Behar at the *New York Times*. But looking a decade out, he told a CNBC reporter a year later, "we should have a number of tiny devices that will be treating conditions we use molecular medicines for today" heralding this "new class of new therapies." All I can tell you is, in tech, beware the ten-year horizon.

After that, electroceutical chat went very, very quiet. (Part of that story concerns a patenting misfire.) Nobody has rice-sized implants routing their nervous signals around the body. Galvani Biosciences is still plugging away, but with replication results that don't make any headlines.

Now, part of this is just the inevitable rollercoaster of the hype cycle. First, you get a big splashy announcement of a new possibility and everyone is very excited. Then the grind of basic research sets in, and there's a long trough of disillusionment because the new hot devices aren't ready immediately. Eventually, positive results start to emerge from the long tail of clinical research, and slowly the once-hyped revolution is integrated into routine care at your doctor's office, and fades into the background of everyday life. And in fact there are signs that this is starting to happen—in 2022 Galvani put the first autoimmune disorder device into a clinical trial.[7]

So maybe electroceuticals are tracing this classic innovation curve. But even after they go through trials, they will face many of the same factors that have been putting the brakes on DBS's ability to perform miracles.

Sticking a pin into the 100,000 fibers of the vagus nerve is unsurprisingly turning out to be much more complicated than initial reports promised, with similar uncertainties and unexpected side effects.[8] Several of these are cataloged in the 2018 book *The Danger Within Us*, written by a former emergency room physician associate named Jeanne Lenzer, who turned to investigative journalism after witnessing the life-altering consequences of the first generation of these implants—these were nothing like the rice grains envisioned by Galvani, but big pacemaker-like devices that had been implanted before there was much understanding of how stimulating the vagus nerve worked to ameliorate symptoms of drug-resistant epilepsy. Lenzer's book shone a particular spotlight on this technique, approved by the FDA long before Kevin Tracey found it might influence immune function. For one of Lenzer's patients, the intervention devastated his cardiac function.[9]

The metal implants we use to stimulate the nervous system simply don't play well with the nervous system.

The trouble with implants

To interact with the electrical signals of the body, either by reading or writing them, you need to use an electrical device. In the brain and heart, implants like pacemakers and deep brain stimulators have traditionally been made from materials deployed in the semiconductor industry, like silicon or metals that control the flow of electricity, among them platinum and gold.

But (unfortunately) your body isn't made of gold. There's no love lost between these kinds of implants and biology, which is likely to mount a healthy campaign of resistance to the invader. That is especially true with brain implants, which create an inflammatory defense response in brains. You can't blame the brain, because during insertion, "the microelectrode tears blood vessels, mechanically damages the membrane of neuronal and [other] cells, and breaches the blood brain barrier," per the authors of a widely cited 2019 study of ways to calm the resulting inflammatory response.[10] Things don't get much better from there.

For people who have no other options—some of whose stories I told in Chapter 5—an electrode can alleviate acute symptoms. But there are trade-offs and problems. For one thing, metals tend to be the wrong thing to stick in a brain. The two materials have a mismatched Young's modulus, which is a way of quantifying a material's "bend or break" quotient. For the brain, the Young's modulus describes not just its give but also its ability to deform and then return to its prior shape. Say you had a bowl of jelly, and you stuck a pencil into it, and carried it around your house. Initially, you'd see no gaps between the jelly and the pencil: they're in perfect contact and the join looks seamless. But after a bit of walking around, you would soon see the jelly de-adhere from the pencil. Worse for the wobbly dessert, apart from the big gaps between jelly and pencil, there would start to be some indirect structural damage in the jelly caused by the destabilizing effects of the intrusion—sideways clefts that split off from the pencil gap. The jelly starts to lose its structural integrity.

You don't want any of this happening to your brain. Once neurons die, they don't regenerate. To try and protect them, the brain relies on supporting characters called glia. They are traditionally considered fighters and janitors that help defend and protect the neurons in their care and keep them running optimally. After an electrode implant, these cells flood in to try to seal the rest of the brain off against the wound made by the rigid, bulky electrode and the dead neurons. To protect the brain's integrity, they envelop the implant in a thick sheath of proteins and cells. This creates a spatial and mechanical barrier that, as it grows thicker, mutes any electrical signals the electrode can both send and receive. Over time, the signals will degrade in crispness and the implant will eventually stop working entirely. At that point it will have to be replaced, which requires yet another brain surgery, another implant, more dead neurons, and more angry glia.

Meanwhile, things aren't going too great for the pencil in our extended metaphor, either. Being sealed off from the electrical signals is not the only trouble for the implant. Biology is hostile to things like silicon and metal. Instead of a harmless tasty flavor, imagine your jelly is a corrosive brine of salt and vinegar. That pencil may look okay for a while, but leave it in the mixture long enough and it will start to take some damage—all right for a £1 pencil, less so for your extremely expensive, sensitive, experimental electrode.

Engineers test implant materials for longevity by bathing the devices in warm salt water for a few weeks to try to recapitulate a couple of years in the environment of a human body.[11] But we have little idea what would happen to an implant you want in your head for thirty years, as the scope for testing is limited; mice only live for, at most, three to five years.

Are you thinking about those so-called telepathic AI-brain implants a little differently now?

There's an enormous research effort to mitigate these problems, with many projects at various stages of maturity. Different rules will apply for neural implants, tissue engineering, and materials for wound healing. But broadly there are two rules, Chris Bettinger tells me, and he knows these rules, because his laboratory at Carnegie Mellon University in Pittsburgh works on creating the materials that need to obey them: "The main ways to make an implant that evades the immune response is either make it very small or camouflage it."

The first rule explains why there has been an enormous effort to make everything at the nanoscale. Tiny wires or grains will be so infinitesimal, so the theory goes, that the brain won't notice the interloper and therefore won't mount an immune response.

The problem with that is that you can only do so much listening or talking with a tiny device. The smaller the electrodes get, the less suited they will be to recording from the brain due to basic physics.[12] You'd have to make up for it by putting in many, many tiny devices. And then your brain *will* probably notice, and you're back to square one with an immune response.

The other option addresses the problem a bit more elegantly: cover the electrical interloper in something the body mistakes for familiar. Many people are trying to come up with a material the body is happy to see hanging out inside its environment and use it to disguise the silicon or metal beneath.[13] The winning material needs to be able to conduct electricity without messing with your brain's structure or otherwise catching the attention of the glia. But what material conducts electrons apart from metals? Well, as it turns out, plastics.

We used to think polymers were insulators—and they are, which is why they are used for, well, insulation. But in 1977, Alan J. Heeger, Alan G. MacDiarmid, and Hideki Shirakawa found out that some kinds of plastic can conduct a current when they discovered a synthetic polymer called polyacetylene. Their fabrication of this "conductive plastic" with metallic-like electroactivity was a major breakthrough in the field—in 2000, it earned the trio the Nobel in chemistry.[14] It's to them we owe the existence of flat-screen TVs and anti-static coatings and all kinds of trappings of modern life. Their discovery also kicked off a new field of research called organic electronics, and twenty-five types of conductive polymers have been developed since.

One major goal for organic electronics has been to solve the Young's constant problem and thereby create ever squishier, more flexible electronics. Some organic semiconductors fit the bill, and one that is getting a lot of attention right now has an unspeakable name typical of the genre: poly(3,4-ethylenedioxythiophene). The material (nickname PEDOT) is so promising it even found its way into the *Independent*: "Scientists have discovered a ground-breaking bio-synthetic material that they claim can be used to merge artificial intelligence with the human brain," they reported in 2020. "The breakthrough is a major step toward integrating electronics with the body to create part human, part robotic 'cyborg' beings."[15]

And PEDOT is indeed nice—squishy, stable, and kinder to cells. But is it going to help *you* become a cyborg? Kip Ludwig, ever jaundiced after his many years in the industry, can contain his enthusiasm: "This is not a game-changer by any stretch." Though PEDOT has been approved for devices like catheters, like the other polymers vying to usher in our cyborg future, PEDOT has some obstacles to overcome before the FDA or other bodies will let you pop it in someone's brain. Yes, it may be the least offensive implant material we have ever made, and yes, it conducts electrons with the best of the rigid metal implants. There's just one problem: we don't speak electron.

Lost in translation

"There's a fundamental asymmetry between the devices that drive our information economy and the tissues in the nervous system," Bettinger told *The Verge* in 2018.[16] "Your cell phone and your computer use electrons and pass them back and forth as the fundamental unit of information. Neurons, though, use ions like sodium and potassium. This matters because, to make a simple analogy, that means you need to translate the language."

"One of the misnomers within the field actually is that I'm injecting current through these electrodes," explains Kip Ludwig. "Not if I'm doing it right, I don't." The electrons that travel down a platinum or titanium wire to the implant never make it into your brain tissue. Instead, they line up on the electrode. This produces a negative charge, which pulls ions from the neurons around it. "If I pull enough ions away from the tissue, I cause voltage-gated ion channels to open," says Ludwig. That can—but doesn't always—make a nerve fire an action potential. Get nerves to fire. That's it—that's your only move.[17]

It may seem counterintuitive: the nervous system runs on action potentials, so why wouldn't it work to just try to write our own action potentials on top of the brain's own ones? The problem is that our attempts to write action potentials can be incredibly ham-fisted, says Ludwig.[18] They don't always do what we think they do. For one thing, our tools are nowhere near precise enough to hit only the exact neurons we are trying to stimulate. So the implant sits in the middle of a bunch of different cells, sweeping up and activating unrelated neurons with its electric field. Remember how I said glia were traditionally considered the brain's janitorial staff? Well, more recently it emerged that they also do some information processing—and our clumsy electrodes will fire them too, to unknown effects. "It's like pulling the stopper on your bathtub and only trying to move one of three toy boats in the bathwater," says Ludwig. And even if we do manage to hit the neurons we're trying to, there's no guarantee that the stimulation is hitting it in the correct location.

To bring electroceuticals into medicine, we really need better techniques to talk to cells. If the electron-to-ion language barrier is an obstacle to talking to neurons, it's an absolute non-starter for cells that don't use action potentials, like the ones that we are trying to target with next-generation electrical interventions, including skin cells, bone cells, and the rest. If we want to control the membrane voltage of cancer cells to coax them back to normal behavior; if we want to nudge the wound current in skin or bone cells; if we want to control the fate of a stem cell—none of that is achievable with our one and only tool of making a nerve fire an action potential. We need a bigger toolkit. Luckily, this is the objective for a fast-growing area of research looking to make devices, computing elements, and wiring that can talk to ions in their native tongue.

Several research groups are working on "mixed conduction," a project whose goal is devices that can speak bioelectricity. It relies heavily on plastics and advanced polymers with long names that often include punctuation and numbers. If the goal is a DBS electrode you can keep in the brain for more than ten years, these materials will need to safely interact with the body's native tissues for much longer than they do now. And that search is far from over.

People are understandably beginning to wonder: why not just skip the middle man and actually make this stuff out of biological materials instead of manufacturing polymers? Why not learn how nature does it?[19]

It's been tried before. In the 1970s, there was a flurry of interest in using coral for bone grafts instead of autografts.[20] Instead of a traumatic double-surgery to harvest the necessary bone tissue from a different part of the body, coral implants acted as a scaffold to let the body's new bone cells grow into and form the new bone. Coral is naturally osteoconductive, which means new bone cells happily slide onto it and find it an agreeable place to proliferate. It's also biodegradable: after the bone grew onto it, the coral was gradually absorbed, metabolized, and then excreted by the body. Steady improvements have produced few inflammatory responses or complications. Now there are several companies growing specialized coral for bone grafts and implants.[21]

After the success of coral, people began to take a closer look at marine sources for biomaterials. This field is now rapidly evolving—thanks to new processing methods which have made it possible to harvest a lot of useful materials from what used to be just marine waste, the last decade has seen an increasing number of biomaterials that originate from marine organisms.[22] These include replacement sources for gelatin (snails), collagen (jellyfish), and keratin (sponges), marine sources of which are plentiful, biocompatible, and biodegradable. And not just inside the body—one reason interest in these has spiked is the effort to move away from polluting synthetic plastic materials.

Apart from all the other benefits of marine-derived dupes, they're also able to conduct an ion current. That was what Marco Rolandi was thinking about in 2010 when he and his colleagues at the University of Washington built a transistor out of a piece of squid.

The squid redux

A transistor is a little piece of silicon in your laptop that can switch on or off the electrical current flowing through it. I don't want to talk about transistors any more than I have to, so please trust me that it is the fundamental unit of modern computing, and that billions of these little guys are stuffed into your laptop and your phone and all your other digital electronics, and they are responsible for the astonishing capabilities of these machines.

Rolandi's transistor looked nothing like the highly sophisticated, exquisitely etched devices that sit in your laptop. It was neither processed nor exquisite, just a few soggy-looking nanofibers of chitosan, a material derived from squid pen, which is a vestigial internal hard bit descended from the animal's ancestral mollusc shell. It's soft and pliable enough that a brain implant would likely cause minimal scarring, but that wasn't its main advantage. The appeal of this transistor was that, unlike the fancy semiconductors that act as gatehouses for electron currents, this one was able to control the flow of protons.

So why are we so excited about protons?

You might remember from Chapter 7, protons are just hydrogen ions. Researchers understand them well, because their contributions to the reactions that make energy in the cell have been studied to death.[23] Protons are also the main component that determines the acidity inside and outside cells. These are among the most thoroughly picked-over mechanisms in biology.[24] So far, so boring, if I can be frank.

Here's something about protons that is not boring: they are able to control the membrane voltage of a cell, and thereby control sodium and potassium and voltage, and thereby cell identity, during regeneration and cancer. "It doesn't matter which ions or ion channels you use as long as you can control the voltage," says Dany Spencer Adams. "What's important is the bioelectric state they create." Protons were the easiest to use. You just had to borrow a gene from yeast to make them. Adams and Levin used this insight to create that mirror-image organ condition in frog embryos.

Controlling the flow of protons would do something that has not been possible—combine the effectiveness of drugs with the local precision of electrical zaps. If you could make an electrical device that could manipulate the proton gradients the way they were altered to make frogs regenerate—but in a more tailored way than a drug could—well, then you'd have a whole new option for bioelectric medicine, a best of both worlds to combine the power of ion-channel drugs and electroceuticals.

Actually, the more you learn about protons, the easier it is to understand what Rolandi found so compelling about a device that could control their flow. If you can manipulate protons in a cell, it becomes possible to do precision tweaks of cellular electricity without involving electrons—or other ions. "It's really easy to use,"

Adams says. "A proton pump is nothing fancy—it's just a single protein." That means it is easy to get into the body. After isolating these proteins from yeast, Adams then simply injected them into the frog embryos. "[The proton pump] assembles itself." That current changed the concentration of protons in the cells, which changed the membrane voltage, which then changed the identity of the cells. Soon, in Adams's experiment, the once-non-regenerating cells agreed to start regenerating again. The reverse was also true: she was able to stop a regenerating frog from being able to do so by poisoning one of its hydrogen pumps to keep it from working. "But it doesn't matter how you inject those protons or control them," she says. "The only thing that matters is the voltage."

In the decade or so since he first made that early smudge of chitosan, Rolandi has honed his device and made many more. And he's not alone. Biological materials from cephalopods are an increasingly attractive research area in general. Chitosan, for example, turns out to be much better at absorbing large amounts of blood than traditional bandages, so it is widely used in wound dressings for military applications.

But it's their electrical properties that have drawn researchers to look more closely at various parts of the squid. The chitosan from its pen conducts not just protons, but other ions as well. A reflective protein in squid skin called—of course—reflectin is also a conductor of protons. Even the ink that jets defensively out of a squid contains eumelanin, which is capable of mixed conduction.[25]

As these properties came to light, more people started tinkering with the materials to see whether they could make a device that could control a non-electron current. Alon Gorodetsky, a chemical engineer at the University of California Irvine, has concluded that reflectin conducts protons fast enough to make it a plausible material for a proton-based protonic transistor—just as the transistor is the basic unit of computation that makes the current flow in your electronic devices, a proton transistor might make ions flow instead.[26] Gorodetsky and his group have also been testing materials from arthropods, and they think these will form the next generation of biocompatible proton-conducting materials and protonic devices.[27] They may even be the basis of edible batteries, which could also be useful for implants.[28]

But for all the advances in the field since his first foray into squidtronics, Rolandi has moved away from cephalopods. "We gravitated toward the biomaterial routes at the beginning," he told me, huffing through an early morning hike near the University of California's idyllic Santa Cruz campus, where he is now the chair of the engineering department. "Back then, my thinking hadn't really crystallized yet." More than a decade after his first foray into biological electronics, he's agnostic about what kind of material he uses. The real prize, he realized, was the ability—by any means—to control protons.

Rolandi started making proton devices to tweak cellular currents out of silver chloride and palladium. The upshot was that protons might be a stopgap until we can figure out how to interface with individual ions and individual channels, and offer a more precise interaction and control than electrons provide. A 2017 paper Rolandi wrote landed on Michael Levin's desk, and he got in touch. He knew exactly what he wanted to do with such a capability.

Levin had found that a cell's fate (muscle, skin, fat, etc.) is tied to its membrane voltage, as we've discussed. Fat cells tended to be around -50 millivolts with respect to the extracellular fluid. Skeletal muscle was the most polarized, at -90. Skin and neurons hovered in the middle around -70. He had also seen that stem cells were nearly at 0, and as their membrane polarized so did their identity develop in accordance with the amount. Now he wanted to tune a stem cell's voltage himself and thereby control its destiny. If you could reliably drive it to become a fat cell, or a bone cell, or a neuron, this would be proof that electricity could be used as a control system for a dizzying amount of genetic and chemical processes.

But how could he keep a living cell in a constant state for enough time that it would differentiate into something new—probably hours, possibly days? The problem with cells is that they are homeostatic—if something perturbs their voltage, they will quickly rebalance. In the body, that problem is solved because the microenvironment around the cell exerts constant regulatory signaling. No existing tool in an electrophysiologist's repertoire was capable of mimicking this.

Then DARPA swooped in to help. It has a long history of investing heavily in research to advance, for example, new directions in prosthetic limbs and neuroprosthetics. Around the time Rolandi met Levin, DARPA also developed an interest in bioelectricity, thanks to the arrival of a new program manager called Paul Sheehan, who had been deeply influenced by Rolandi's proton transistor. (In the course of a previous appointment at the US Naval Research Laboratory, Sheehan had used proton pumps to design color-changing bioelectronic devices, based on squid camouflage.[29])

Now that he had a purse at DARPA, Sheehan gave Rolandi and Levin money for their stem cell fate project. With the money, Rolandi and Levin brought Marcella Gomez onboard. Gomez is a mathematical and systems biologist at Santa Cruz with a background in control theory and cybernetics. She knows her way around mathematical tools that can drive biology and realized what they needed was a machine learning system that could monitor the constantly changing voltage of the cells and act on it in real time. So she made one.

The team put stem cells into an array with a device of Rolandi's design, which injected a proton current around the cell to drive its membrane voltage up. Whenever a cell would engage any of its channels to try to get back to a more comfortable voltage, Gomez's AI would notice and inject more proton current. It was able to consistently keep the membrane voltage of living stem cells 10 millivolts higher than the cell's usual depolarized baseline. In 2020, the trio published the results of Gomez's remarkable new tool, which had managed to continuously impose the artificial voltage for ten hours. No one had ever done that before.

As they were working out how to extend that voltage window so that they could watch the stem cell differentiate, however, funding ran out.

But that turned out to be okay, because by then, they had given Sheehan all the evidence he needed to launch the much bigger project he had been planning. At the beginning of 2020, DARPA launched the $16 million BETR (Bioelectronics for Tissue Regeneration) program—a pretty hefty chunk of money even by DARPA standards—whose goal is a radically expedited healing process for wounds.[30] This has not been possible with traditional electronics (or anything else, for that matter). While individual studies on electrical stimulation for healing have sometimes had promising results, no one could ever give you a specific recipe for how to make it work every time in every patient. Sheehan had seen enough of the research to suspect that speaking to the body in its own language might be the way out of the impasse. "I wanted to pivot to bioelectricity mediated by ions, instead of just a voltage," he told me. "Right now, it is very challenging to go from electrical to biochemical signals and vice versa. That's what this program is trying to do." He wants to improve every aspect of wound healing, from better sensors and actuators to creating better models of how healing actually works.

There are so many things we don't know about wounds, and this is why no one has come up with advances to make them heal better or faster. One problem is that every wound is different. Sheehan ticked off the list for me. "The edge of a wound is different from the center. A cut on your foot heals at a different rate than a cut on your face. Young people heal faster than old people."

Rolandi's group is using bioelectronics to control different aspects of wound regeneration. Instead of just applying an electric field and hoping for general improvements, the idea is to be specific. The team monitors specific wound processes (like the inflammation stage) with sensors. Gomez's algorithms then process information from these sensors into actionable items, for example to deliver ions or an electric field to the wound to calm macrophages faster in a bid to accelerate the healing process. They couldn't get that granular without more diverse tools at their disposal. "Otherwise it would be like, great, you've detected all this stuff, and you have this very complicated algorithm, and now all you can do with the information is shoot an electron at it," says Rolandi. "That's just not going to cut it."

But just as the stem cell project was for Sheehan a stepping stone to the BETR project, so is BETR a step to something bigger. "Wound healing is a great initial problem to go after," he said. "But if you look across the board, there are many different places in medicine where you want to have control over the delivery of a drug compound." One oft-cited example is delivering specifically targeted medicines to a tumor, but a generic interface could also choose the time of delivery not just the place. Any oncologist will tell you that they wish cancer medications could be administered to their patients at night when they're asleep, because that is when the body is regenerating. What's more, during this resting period, some non-cancerous tissues that are most sensitive to drugs are not dividing—administering noxious drugs then would help reduce some of the adverse consequences. But of course you can't administer these drugs in the middle of the night. The doctors and nurses and office managers are asleep too.

Hence Sheehan's next goal: "What we really need is a generic interface with biology that would enable us to deliver biological information into the body," he told me. "Cytokines, hormones, chemokines. Having a generic device that could deliver those therapies would be like having a twenty-four-hour doctor right there," says Sheehan. Or for wounds, it would act as a twenty-four-hour surgeon on call. As it happens, that was one of the selling points of the xenobots.

Frog robots and fungus computers

When Michael Levin first started dismantling the frog embryos, he wanted to understand what happens to living cells when they are freed from the constraints of the electrical signals sent by their bioelectric environment. You remember from Chapter 7 that he and a coterie of other scientists believe that these cues are a crucial authority that instructs cells which shape to assume and where, and that this guidance is crucial to whether those cells cooperate in their trillions to properly form us in the womb.

But how do you put that idea to the test? "The xenobots were a way of asking: here are a bunch of cells, how do they specify what they should be building, in the absence of any guidance?" Levin told me. "The big picture is not about having robots made of frog cells, or in fact, robots made of any cells. The idea here is that we need to understand how collections of competent agents work together toward large goals." That would have obvious corollaries in regenerative medicine: how do cells get together and agree on building something large, like an organ, or in fact, the whole body? It might also yield insights into why and under what circumstances cells opt for an "every man for himself" approach to become cancerous.

"Everything in my lab focuses on this idea of how many become one," he says. "How is it that lots of small competent agents get together to have a single unified cognitive system that has a goal state?" If we understand how that works, then rebuilding organs, reprogramming tumors, fixing birth defects, and reversing aging really does just become a matter of programming. "Everything boils down to the question of convincing cells to build one thing rather than whatever they're currently doing," he says.

So he decided to see what the cells would do in the absence of the cues. He and his collaborators scraped a few thousand cells off a frog embryo. Then they put them into a completely different, neutral environment and waited to see what they would make of their newfound independence. The cells had a lot of options. They could have just died. All the cells could have struck out on their own. They could have formed themselves into a single layered "skin" that lay in a flat plane like a cell culture.

They did none of that.

Instead, a few thousand of them got together and made something new. Somehow, they agreed among themselves to glom together into a new architecture, little discrete balls. Then each of them grew cilia, which in itself wasn't unusual. These little hairs grow on the outer surface of normally developing embryos to move the mucus around the body and keep it clean—what was unusual was how they put them to use. "These cells basically repurposed that genetically encoded hardware," Levin says. Now, instead of using their cilia to move the mucus, they used them to move themselves—even though they didn't have any nervous system to either conjure intent or act on that intent. Nonetheless, with their new equipment in place, they started coasting around.

"We have these amazing videos of the little clumps moving around. Sometimes they formed little groups, interacted in various configurations, they even went through a maze."

And even though they were just clumps of cells with no brain or nervous system, they even seemed to have *preferences*. When Levin cut them nearly in half, they regenerated, always seeming to favor reconstituting themselves into the little spherical shapes they had originally assumed, if a ball of 2,000 frog cells can be said to prefer something. Xenobots indeed. "These are neither a traditional robot nor a known species of animal. It's a new class of artifact: a living, programmable organism," said Joshua Bongard, the roboticist on the team.

So far, the only thing that is strictly programmable about them is their shape and their lifespan. The xenobots don't have digestive systems and so their cells are instantiated with a little yolk sac containing a fixed amount of fuel. When that runs out, they die. That does seem to be the main advantage of using living systems as robots—living systems die, which rules out the horror scenario of xenobots overrunning the world.

Or maybe not. By the end of 2021, they had been programmed to reproduce.[31] They had not built themselves novel sex organs—instead, they used their Pacman-like mouths to scoop up cells into groups of roughly the same size as themselves, which then themselves aggregated into new lifeforms. They made new creatures like themselves. It was a method of reproduction that was new to the evolutionary history of the planet. Nearly five years into working on the creatures, Levin is unequivocal: they are alive—"under any reasonable definition of life." No wonder ethicists began to worry. "Is this as much of a Pandora's Box as it sounds?" wrote two of them not long after the self-reproduction study was published, raising a range of possible adverse consequences and wondering whether science needed more limits placed on it to avoid them.³² "Although xenobots are not currently made from human embryos or stem cells, it is conceivable they could be," they wrote.

Andrew Adamatzky thinks biology is the inevitable future of implants, but where others are working with frogs and squids, his money is on fungus. Adamatzky is a professor of unconventional computing at the University of the West of England, where he has created a computer model of the mycelial electrical activity and encoded the spikes into logical functions, a bit like AND/OR functions transistors are able to create in traditional computing.³³ Once we have these for the body, why not for the environment?

The future is not diving to the bottom of a reef to source some coral for your hip. The future is to understand the properties of biomaterials that make them good interfaces, and then make a steady supply of them, tuned to the properties that can best interface with the body—synthetic coral, synthetic squid pen, to ensure a steady supply of materials whose quality is as unerring as the silicon crystals that now make semiconductor wafers.

While we wait for the new ion-channel drugs, the new trials, and the biological implants (none of which are guaranteed to be with us in ten years), there is another option for electroceuticals: non-invasive wearables that can do it all from outside the skin.

CHAPTER 10

Electrifying ourselves better: New brains and bodies through electrochemistry

Mike Weisend plucked two bespoke electrodes from their nest of protective foam: these were the large daisy-shaped discs that would channel the electricity through my brain. He asked me to hold one to my right temple as he strapped it to my head with gauze. Then he squirted a large gloop of the green liquid into the cut-outs. He explained that the daisy at my temple—and the other at my arm—would pass a harmless amount of current through my skull.

We walked into a windowless gray office that the lab's decorators had done their level best to turn into a military theater of operations. At one end was a mound of sandbags, piled up about shoulder height. Resting against it was a big M4 rifle: a model often used for close combat. I shouldered it. Against a wall about ten feet in front of the sandbags was a projection of a training simulation called *DARWARS Ambush!*

I was there to try an experimental technology called transcranial direct current stimulation (tDCS for short). I had first encountered the idea at a military conference put on by DARPA, the division of the US military that has birthed world-changing breakthrough technology like the internet, GPS, and lasers. (Their conference has since been discontinued, probably because it allowed nosy journalists like me to find scientists who were accelerating soldiers' learning by zapping their brains with an electrical current.) There, I had learned about a new technique they were using to speed up sniper training, using bursts of electricity administered to the cranium. They were guarding this program so closely that it took four years of pleading with DARPA before they would agree to so much as a twenty-minute follow-up call. Little wonder, given their results: "For soldiers who are learning marksmanship, we've been able to cut the time it takes to get them from novice to expert in half," the program manager told me on that phone call. They had achieved similar results for languages and physics.

Did I really want to take these remarkable results at their word? What I really needed was someone who could tell me what it was like, and they wouldn't let me meet any of the soldiers from the trials. "Can I try it?" I ventured.

A short pause, and then there was an intake of breath as if he was about to talk. "I'll sign any waiver you need me to sign," I pre-empted, already intoxicated by visions of my own electrically mediated virtuosity.

Another pause, this one longer and definitely with the speakerphone muted. "You'd need to come to California—"

"Okay!" I said, before he could finish the sentence.

About a month later, I was on my way to the US west coast. In the fast run-up to this experiment, and my general excitement, I had made some poor judgment calls. The first was scheduling the meeting the morning after an eleven-hour flight from London to California, at the wrong time and in the wrong direction for sleep. Then there was the drive up and down the mountains where I had decided to stay with a friend to save *New Scientist* the hundred dollars for the hotel room. It turns out the LA mountains are a lot higher that you'd expect. Thanks to jetlag and altitude sickness,

I hadn't managed to sleep in more than thirty-minute chunks since I'd boarded the plane. Fueled by dangerous amounts of coffee, I drove down the sickening decline in the pre-dawn blackness, crying a little bit and shrilly repeating the mantra, "Great, I guess this is how I die!" And that was before I hit the traffic.

By the time I arrived to meet the team, I was too busy berating myself to consider how I was going to approach the challenge that lay ahead. Four years of chasing this story across two jobs, a transatlantic and transcontinental flight—and I hadn't bothered to allow any time to prepare my brain for a neuroscience experiment? I would have got a better story just by transcribing the DARPA interview I had conducted from my desk chair in London. I quaked silently with self-directed rage.

Michael Weisend's waist-length fall of salt-and-pepper hair didn't increase my confidence. Weisend, a neuroscientist who at the time was with the University of New Mexico, had been kind enough to fly to California that morning to demonstrate his electrical apparatus. He took me into a small room where I found a bulky suitcase whose foam-padded insides nestled around an assortment of wires, a squeezy bottle full of ominous neon green liquid, and a beige box festooned with switches and dials, which housed a 9-volt battery. Weisend chortled as he unpacked the ingredients. "Can you imagine getting this stuff past airport security?"

After Weisend had finished attaching the electrodes to my body, he tucked the chunky rig into the back of my bra. "You're all set," he said. It was time to go to war.

It started easily enough, with some electricity-free target practice, during which I familiarized myself with the weight and heft of the modified gun. I found myself in a simulated desert, no sound but whistling wind, facing a row of roughly humanoid metal targets. Whenever I hit one, the bullet ricocheted off with a satisfyingly realistic *ping*. A series of these scenes followed. I did all right despite the fatigue.

Weisend came back in. "Okay, now we're going to see if we can make this as realistic as possible," he said, fiddling with the box behind me—he meant he was going to try to approximate the control and sham condition on a clinical trial. That would require me having no idea whether or not the electricity was on, so that I wouldn't get messed around by the placebo effect. "I'm going to come in a bunch of times, but I'm not going to tell you when I turn the electricity on." This wouldn't have passed a review, but then I wasn't actually participating in a clinical trial. This was an anecdote, and I was a tourist.

He left, and the quiet dunes and targets dissolved. I became a sniper at a checkpoint. More specifically: I became a *terrible* sniper. I was a nervous wreck even before anything happened. My eyes flickered back and forth manically between the building and the incoming cars. Any minute now something would happen, but I didn't know what.

It was almost a relief when the bomb went off. As the white blast faded, a man ran at me wearing an explosive vest. You may recall the rest from the introduction of this book. The simulation faded out in a gray mist.

The tech came in and reset my rifle, and I re-spawned once more into the checkpoint. This time I knew what was coming, and was prepared for the first bomber. I also managed to dispatch the shooters on the roof, but after the second bomber came at me, suddenly dozens came running impossibly fast from several directions at once. Gray mist.

I can't remember how many more times I went through it—three? Twenty? All I know is that every session seemed interminable, and by the time the lights came up the last time, I just wanted it to stop.

I was also beginning to wonder if the whole thing was a scam. Recently published experiments showed tDCS-enhanced training increased a sniper's ability to detect a threat by a factor of 2.3, but I wasn't seeing any of those results. After all, there is a long history of defense contractors in the US overinterpreting—or sometimes downright falsifying—their research for eager government procurement officers. I was getting steadily more resentful, dreading the drive back through the traffic, reeling with fatigue.

Weisend came in and fiddled with the device again. I tasted metal, a bit like I'd just licked the tab on an aluminium can. This was it—even though I wasn't supposed to know the sham experiment from the real, my permanent dental retainer had given it away. Despite my earlier skepticism, I was suddenly excited. I waited for my *Matrix* moment. Any minute, new information would rain into my mind like 1990s-movie code hieroglyphics, filling me with the sudden ability to comprehend the physics of shooting. But that still didn't happen. There was only the metallic taste. I sighed heavily and resigned myself to another humiliating in-game death.

"I'll see you in a little while," Weisend said, and left. The lights went down again. And without much fanfare, I calmly dispatched all comers in a session that felt like it took three minutes, though Weisend (and the tech, and several wall clocks) would assure me that it was twenty.

"How many did I get?" I asked the tech when the lights came up. You know the rest.

The question I began asking myself back then, and that has continued to drive me since: how is it possible that an electrical current that lights up your laptop can manipulate the delicate natural electricity that makes the body run, to such staggering effect? And when can I get a device of my own? And should any of us be able to do this?

After the experiment, I developed a particular obsession with question number two. I remember being at a work social function a few months later, and surprised myself by welling up as I recounted the experience to one of my colleagues. It went beyond the experience in the lab itself. It was the drive back from the lab—I wove calmly in and out of traffic, and driving was so pleasant, where it's more often a white-knuckle gritted-teeth affair for me. It was the following three days, over the course of which I approached incoming problems exactly the way I had approached those fake assailants: calmly, without panic, without the complicated ritual where I first needed to dredge up the lifelong list of my own failures and genuflect before my own worthlessness. That bottomless font had suddenly dried up. And that meant all of a sudden, life was so much *easier*. Who knew you could just, like, *do stuff* without first performing the elaborate dance of psychological self-recrimination?

And how the fuck, excuse my language, had a little electrical zap done all of that?

One theory had been that this was a non-invasive way to boost alpha oscillations. These, as you might recall from Chapter 5, were discovered by Hans Berger. For the better part of a century, the oscillations he identified were thought to be epiphenomena, mere "exhaust fumes" of the brain: they could tell you simple things like whether the engine was running, and even sometimes dispensed limited information about its condition. For example, in the 1930s studying oscillations with EEG helped Alfred Loomis advance the study of sleep science. The now common idea that sleep happens in stages like REM and non-REM would be inconceivable without these different telltale waveforms to tell them apart.

A handful of animal experiments had suggested you could technically alter brainwaves but without the invasive precision of a penetrating implant, you couldn't target specific functions, and there wasn't a use case for doing so in a human, even if you could get approval.

That all changed in 2000, when two neurologists at the University of Göttingen in Germany, Walter Paulus and Michael Nitsche, published a paper describing a new technique called transcranial direct current stimulation. With tDCS, it was possible to alter the rhythms of the oscillations without brain surgery—and see if changing the rhythms would change a person's behavior or mental state. It was relatively easy and safe: strap two electrodes to a volunteer's head, position them over the brain areas of interest, and set a very mild current flowing (anywhere between 1 and 2 milliamps). In 2003, Paulus's team published an experiment that seemed to show that tDCS could boost cognitive performance, accelerating people's ability to learn a random sequence of keystrokes on a computer keyboard.[1] "It was like giving a small cup of coffee to a relatively focal part of your brain," one of his co-authors told *New Scientist*.[2]

That's when tDCS exploded. Now everyone was on the hunt for ways to improve the brain with this easy new tool. A year later, Lisa Marshall at the University of Lübeck sharpened people's memories by boosting the size of a kind of squiggle called a sleep spindle with short bursts of tDCS, as they slept.[3] The next morning, they could better recall word pairs they learned the previous day than people who hadn't had their brains sleep-zapped. Other researchers jumped to replicate this and other mental boosts. At universities like Oxford, Harvard, and Charité, a little bit of electricity enhanced memory, mathematical skills, attention and focus, and creativity. By 2010, thousands of papers had been published, purporting to show the effects of electrification on memory and cognition.

Here's the problem. It did not do that for everyone. It didn't even reliably do it for me. As I mentioned in the introduction, while tDCS worked a treat for sharp-shooting, it didn't touch my mathematical deficiencies.

Extraordinary claims require extraordinary proof. It began to emerge that many of the studies whose dramatic results had been so widely reported didn't even have ordinary proof. They had been drastically underpowered, with participant numbers straining credulity in the low single digits. Some had no control group at all—science's mortal sin. It wasn't just the badly done science that was posing a problem for tDCS. Even the good studies were under siege, because there was no firm consensus about how exactly tDCS was supposed to be creating all these effects. Meanwhile, a number of the people who had purchased one of the many new home tDCS kits began to complain that they had no effect.

Then a couple of studies were published that raised the question of whether tDCS was a giant scam. In one somewhat gruesome experiment, New York University researchers tested the effects of the standard tDCS dose—the same 2 milliamps I had experienced in California—on a cadaver. That dose didn't even send enough electricity through the skull to show up in the brain, they said: 90 percent dribbled away into other parts of the body, including the skin of the scalp. How could something like that have any effects on cognition? Even the researcher who had done the DARPA protocol on me was sympathetic to the naysayers. "For every good study, there are an equal number that throw shit at a wall to see if it sticks," he told me.

Indeed, by 2016, what had started out as a way to probe the functional role of oscillations had turned into a full-blown putative panacea. You name it, someone had got a grant to zap it out of someone. "Here is a list of everything tDCS is supposed to work for," declared Vincent Walsh of the University College London's

Institute of Cognitive Neuroscience at a tDCS summit, before enumerating a list that included schizophrenia, eating disorders, depression, migraine, epilepsy, pain, MS, addictive behavior, poor reasoning, and autism, and stretched from there into the double digits.[4] "Do me a favor . . ." he said, voice dripping with the finest British vinegar. He was certainly not alone in being reminded of the post-Galvani era of electroquackery.

This was happening because it had not taken long for the focus to shift from oscillations to the tool to manipulate them—and somewhere along the way, the oscillations themselves got lost in the shuffle. Silicon Valley, too, had interest in "overclocking" the brain and funded development into alpha wave–boosting technologies—and a tidal wave of gadgets emerged for home use, none of which seemed to actually work. The overfocus on tDCS (Did it work? Did it change the brain? Were action potentials being generated?) obscured the point of using the tool in the first place, which was to see if brain-wide oscillations—not individual action potentials in individual brain areas—could be altered to have behavioral consequences.

As the tDCS controversy died down, other approaches to boosting alpha oscillations renewed the focus on oscillations and whether they were functional. Transcranial magnetic stimulation (being zapped by a giant magnet), deep brain stimulation, and transcranial alternating current stimulation (which wasn't galvanic current but a fast series of pulses that switch really quickly from negative to positive current flow) painted a vast new picture about oscillations: not only could they tell you deep realities about what was going on in the brain, but changing them could change the related behavior.

For all his vinegar, Walsh is not even a kneejerk tDCS skeptic—he has contributed his own studies to the canon. What made him (and Kip Ludwig, and many others) so grumpy about tDCS is how credulously small studies were being reported to, and in, the press. Some of them were barely more credible than my (completely unreliable) gonzo stunt. And yet, people didn't get to read about the fact that the studies didn't have proper control and only five subjects. They read about their promise and saw the device was non-invasive and thought that was equivalent to not having any risks. So, many of them decided to make one for themselves. Reddit has a board dedicated to overclocking your brain, where they offer circuit diagrams and other instructions. I sympathize. Full disclosure, I ended up buying a brain stimulator myself (I don't have the talent to make one). I still couldn't tell you conclusively whether it's placebo or not. I only use it when my brain comes at me with the List.

Lucky me—some people who built their own rigs suffered dire consequences, blinding and burning themselves trying to replicate the exact parameters that would effectively stimulate their brains. Enough that a group of neuroscientists published an open letter begging them to stop.[5]

Déjà vu all over again

Over the past few years, more tDCS studies have proliferated. Like all bioelectric treatments, it works or doesn't based on the tiniest and most unpredictable factors. There are dozens of variables to account for when designing the experiments. You even have to account for variability in skull thickness! (Insert joke here.) Some people get lucky and happen to have the right kind of parameters to go with the electrical stimulation they're given.

That seems to have been the case for me, as I inferred several years later after a chance conversation with Camilla Nord, who studies the effects of tDCS on depression. When I told her that my negative self-talk had been cleared away like so much San Francisco morning fog by the electricity, she lit up. She said she had identified a population of depressed people whose disease manifests in this exact kind of castigating self-talk, where you just spend all your energy on dragging yourself. Their symptoms were particularly lifted by the intervention.[6] But like Helen Mayberg's deep brain stimulation quest in Chapter 5, she was still figuring out how to discern responders from non-responders.

The study of brain stimulation is not broken—it's just really, really hard, like science in general.[7] There's no conspiracy in science to get bad results past peer reviewers; it's just that any one single study can run into loads of problems—not enough funding for sufficient numbers of trial participants, researcher bias, non-standardized equipment, stimulation strengths—there are so many parameters, you hardly know where to start.

But a clinical trial always has to start with some low number of patients (so as to use few resources, which are supposed to be saved for the big final trial). That's standard. Their size makes them more prone to bias, though. That doesn't mean the early data from small trials is worthless, former NIH director Kip Ludwig explains—it's eventually supposed to feed into the big definitive study that can tell you a big definitive result.

The trouble is we've forgotten that those early studies aren't evidence—all the shortcomings I mentioned above lead to a statistical scenario in which you are quite likely to get a "false positive" that your intervention works great (when in reality it doesn't). This is unfortunately what wound up happening with Ivermectin and Hydroxychloroquine to treat Covid, where people who aren't scientists put too much stock into an early study or two, with too few patients and flawed experimental designs. The later, more definitive, more resource-intensive studies later revealed those early results to be a fluke. But by then the misinformation was out there.

We might go through another round of this game of promise-and-recrimination quite soon. Like tDCS, electroceuticals have gone non-invasive. The therapy is now called "vagus nerve stimulation technology," or VNS. It's getting a large infusion of cash from Silicon Valley and is all over social media, but hasn't quite taken the form that was predicted ten years ago. Instead of supporting penetrating implants, many investors are backing non-invasive wearable technologies that try to affect the nerve from atop unbroken skin, such as little earbuds that stimulate the vagus as it climbs from the depths of the body to nearly breach the surface of the skin at a spot just inside the ear. Once again, it's being mooted to help with focus, anxiety, depression . . . By now you know the rest. Similar to tDCS, for each study that shows a hint of an effect in a small number of patients—some of which aren't very well done—there is another study showing it doesn't work.[8]

If we want to understand our electrome well enough to manipulate it precisely with noninvasive gadgets, the first step is huge trials with invasive technology that can conclusively show how the technology interfaces with our bioelectricity.

This raises the question: who is going to let you open their brain to get that data? Every tool and scrap of knowledge we have assembled about the body's electrical dimensions so far has been given by people for whom enrolling in one of these studies is an option of last resort—from Catharina Serafin's heartbeat to Matt Nagle's work with BrainGate to VNS pioneers. Cancer cures, limb regrowth, reversing birth defects, neural upgrades, immune modulation—that healthy people will use to upgrade themselves in the future—rest on the next generation of test pilots.

The test pilots

Around the time she was trying to steer the FDA onto the right track with the oscillating field stimulator, Jennifer French founded Neurotech Network, a neurotechnology advocacy group that helps people with neurological injuries find assistive technology that can help them in their specific circumstance. "Technology can become a great equalizer," she says. "It gives people choices."

However, often people who design neurotechnologies can focus on the kind of evidence that will moisten the eyes of people who are like them, at the expense of things that will help the actual people who need the technology. French understands why they do it. "Getting people walking again is sexy," she says. Behind the scenes, after the media attention has died down, the researchers will quietly file the grants that contain the sorts of priorities that are actually important to people who have a spinal injury: pain, and bowel and bladder control. "Addressing the true needs of this population doesn't sell in the media."

Instead, the public message remains on what the disability studies scholar Stella Young calls "inspiration porn," and it has wide-reaching consequences.[9]

For example, those research videos of paralyzed people walking make the rounds on social media and in the traditional media. Shorn of their context—which they often are, because you know how the internet works—they give people a deeply skewed picture of what is possible if they get injured.

"Every time one of those news stories comes out—*Hey, we've cured spinal cord injury, we have people up and walking!*—that leaves a false impression," French says. "Then the advocacy groups get tons of phone calls from people living with the condition, asking 'When can I get the cure?'" It is not of course a cure, and it falls to groups like hers to bring people down from the hype. The media hype is destructive in more ways than one.

The misperception it seeds also makes it hard for people to have a clear picture of what capabilities actually exist. This makes it difficult to objectively evaluate whether to enrol in a trial.

When Phil Kennedy underwent voluntary, extremely risky (and potentially unethical) surgery on himself, he was widely hailed in the tech press as a self-sacrificing hero of science. And yet, with a few exceptions, the coverage of the volunteers in clinical trials has little of this veneration. "The people who test neurotechnology are test pilots every bit as much as Chuck Yeager or Buzz Aldrin," French says. Just as those men risked their lives to expand science's understanding of the sound barrier and space flight, people who volunteer to test new neurotechnologies should be understood as daredevils working—at great risk to themselves—to take science into new frontiers.

Yeager and Aldrin (and Kennedy) knew the risks inside out before they undertook their experimental flights. But there are no standards for how clinicians should communicate expectations to people who volunteer for trials. As many people join trials of new invasive experimental neurotechnology for altruistic reasons as they do hoping this trial ends up being the cure or an aid. Sometimes a volunteer comes into a trial desperate, their head filled with misleading ideas from the "inspiration porn." French is exasperated by this because the people who join clinical trials are not guinea pigs, and should not be condescended to or tantalized with false hope.

Making someone a test pilot is only ethically possible when that person has the full picture of everything that could go wrong, and precisely managed expectations of what the technology can and can't do. Right now, that kind of transparency is not enforced. "We need to be really clear with people about what this can do for them personally," French says. But there are no standards clinicians must adhere to when advising their trial volunteers.

For one thing, anyone who designs bioelectric interfaces should look at the ethical history of medical implants. We know about the grisly history of people having implants against their will—but what about explants? A few people had their experimental implants removed against their will after the companies that made them went broke. I spent a few hours at a neuroethics conference a couple of years ago talking through these issues with Frederic Gilbert, a neuroscientist and philosopher at the University of Tasmania who studies explantation.

Gilbert points in particular to a major ethical issue—potential trial participants are often not told the whole story about the future of their devices. A study from Rice University and Baylor College of Medicine found that potential study participants are generally soft-shoed when it comes to what will happen to their implants after a trial has concluded.[10]

A typical case would involve a person with a treatment-resistant disease that was robbing her of her quality of life. Maybe she could no longer drive or work. As a last resort, she would join a clinical trial of an implant that promised to change all of that. The implant would work. Soon she could drive, make plans, and regain the fairly predictable life most of us take for granted.

But her implant was an experimental device, and when the neurotech start-up that had implanted her found their device did not work for everyone in the trial, the company collapsed into bankruptcy. The insolvent company could no longer support their devices, so they needed them back. That meant another brain surgery to remove the investigational device. She was unprepared to return to life before the implant. She did not consent to having the device removed, or to the brain surgery. "How are you supposed to recover these devices?" Gilbert asked me. "Do you hunt these people down? It becomes like something out of *Blade Runner*."[11]

When radical new medical technologies are successful, the tech press reports breathlessly on paralyzed people who can now pick up grapes, or locked-in people able to send their thoughts via brain implant. But what happens when the trials are over? Too often that miraculous technology either comes out or becomes inert.

You might be wondering: why can't these devices stay in? Usually it's because they need long-term tech support a failed start-up won't be able to provide. Stimulator batteries need to be changed, or stimulation frequencies adjusted. There needs to be someone in charge of medical check-ups for people with implants bedded down in their gray matter. In rare cases, this is possible—if the person in charge of your clinical trial happens to be Helen Mayberg. You could call Mayberg the doyenne of DBS—after a long and prestigious career at Emory University, the Icahn School of Medicine at New York's Mount Sinai Hospital created a new Center for Advanced Circuit Therapeutics just so she could run it. When you implant patients, she says, you own them. "Not in the sense that you can do what you want with them—it's the opposite," she says. You now bear a lifelong responsibility to them. Mayberg is passionate about this and has clawed her way to letting her participants keep their depression-busting DBS implants after a trial. But she is also a big, powerful name in neuroscience with a lot of credentials to back her up, institutional firepower at her behest, and university funding.

Hank Greely, a law professor at Stanford and an expert on ethics in biosciences, thinks the answer is that before any neuro-engineering or bioelectricity researchers run any kind of trial, the companies or universities behind them should be forced to invest in a bond. "Some kind of common fund which will allow people to keep their devices, have them maintained and repaired, and their batteries replaced," he says. "These people are not rats. You don't implant them, grab your data, and get rid of them."

Today, French lends her expertise to several neuroethics panels and patient advocacy panels, including the National Institutes of Health, the BRAIN initiative, and the Institute for Electrical and Electronics Engineers, which is working on a framework for neuroethics around medical devices and neurotechnology devices. All the new standards aim to ensure full disclosure for the volunteers who are trialing deep brain stimulation, spinal stimulators, and other next-generation neurotech. This is part of the larger neurorights initiative that is gaining traction, most recently having been written into law in Chile in 2021.[12]

Don't mess with the electrics

More volunteers, who are better informed, will accelerate our understanding of neural implants, electroceuticals, and other kinds of electrical intervention. But electrical stimulation is not the only way we are impacting our normal bioelectric functions.

We're starting to look for future electro-drugs in the ion-channel drugs we have been using for decades. They are ion-channel manipulators, capable of either blocking them, prying them open, or otherwise messing with their state. As we saw in Chapters 7 and 8, a more complete understanding of their significance in bioelectric signaling is driving new research into how these drugs could be repurposed for cancer therapies and regenerative medicine. But it also opens up a more unsettling question: if we're already taking so many of these drugs, do we have a complete understanding of what they have been doing to our electrome? Should we start figuring it out?

We started using drugs that acted on our ion channels long before we actually knew about ion channels. We used them because they worked—we figured out *how* they worked later.

In some, bioelectric side effects are already well understood. Most epilepsy drugs, for example, are acknowledged to cause a range of birth abnormalities if taken during pregnancy. As it turns out, this is because of how they muck with our bioelectricity. Many of them suppress overactive sodium or calcium channels, but while this helps to calm the relevant neurons and stop seizures, increasing evidence suggests it may also disrupt the ion-channel communications necessary to correctly pattern a fetal structure. In one medication, the severity of the potential consequences—a significant risk of lifelong learning and cognitive disabilities, and physical anomalies—have led to restricting the medicine's use in people who can get pregnant, during peak reproductive age.

Epileptic drugs are far from the only ones with wide-ranging effects on ion channels—and yet there has been little research into how others might mess with the complex ways ion channels are involved in development. Like Kip Ludwig, Emily Bates works on the nitty-gritty details of bioelectricity, but she does it as a developmental biologist at the University of Colorado School of Medicine. Bates had long wondered what other drugs can screw up the ion channels and thereby result in birth defects.

A quick caveat before I go on. It's very, very early days for some of this research. Talk of medical effects on development is often tinged with a particular strain of prim authoritarianism. When you're pregnant, you aren't "allowed" to do much of anything without conjuring stern words of concern. I would be mortified if my book were used as yet another cudgel to shame someone at a point in their lives when everything is already quite unsettling enough. This is why it is so important to fund research that lets us know what is safe for a developing fetus.

Bates decided to focus on drugs for which a robust body of research already existed, to demonstrate their ill effects on pregnancy. Smoking, for example, has been widely shown to "increase the risk of health problems for developing babies, including preterm birth, and low birth weight," confirms the Centers for Disease Control, along with being firmly linked to birth defects of the mouth and lip, like cleft palate. But it has been hard to tease out which of the 7,000 ingredients in cigarettes is the culprit, as they contain such a vast assortment of chemicals—including ammonia and lead—many of which are implicated in cancer. This is part of the reason vaping has been quietly accepted as a harm reduction campaign—it's the nicotine without all the rest of the stuff.[13] Maybe nicotine in this form is still not great, and it's not quite doctor-recommended, but it's no shock that smokers will often switch to vaping when they become pregnant. And if they vape already, they may not try to quit when they find they are pregnant. Either way, vaping nicotine often results in a higher dose.[14]

Does fetal nicotine exposure cause birth defects? Bates exposed pregnant mice to pure nicotine by putting them into a vaping chamber (essentially a giant walk-in bong) and found that the newborns still had several characteristic developmental problems: they had shorter bones, notably of the humerus and femurs (correlated with shorter stature in humans), and the nicotine exposure had altered their lung development.[15] So nicotine was not the innocent bystander here, as it was clearly causing these defects. So vapes, which provide nicotine, are bad for a developing baby.

It's not yet possible to neatly tie these physical effects to a mechanism. But there's a lot of other research whose pieces fit intriguingly well with the new data to form a compelling picture. It's well established, for example, that nicotine binds and blocks a potassium channel that's known as "inwardly rectifying"—that is, it keeps the concentrations in the cell at the cell's "happy place," because the potassium channel's job is to allow more potassium ions to enter the cell than leave the cell. Bates has spent her career on this one channel. Alcohol, it is turning out from early evidence in her lab, may affect it as well, which may be a culprit for the birth defects associated with fetal alcohol syndrome.

Anesthesia also does weird things to ion channels in ways we don't quite understand yet. It's not just during pregnancy that drugs might affect bioelectric signaling. If you've ever had general anesthesia, there's a small chance you may be at higher risk of developing cancer later in life,[16] or maybe memory problems.[17] There are even cases where people appear to be under but are not, or present lingering, mysterious PTSD-like symptoms.[18] But we don't know, because we actually don't know exactly how anesthesia does what it does. Well, we do know some things. "We know what it does to neurons," says Patrick Purdon, a professor of anesthesia at Harvard. It makes them fire in ways that are completely different from normal physiological processes, in some cases shutting down all firing entirely for seconds at a time. The result of this is that complete gone-ness, more complete than any sleep. We might know the neurons stop—but we don't have a molecular explanation for how they do.

Or to be blunt, how they turn back on again. "The amazing thing about general anesthesia is that any of us ever come back from it the same person as who went in," says Michael Levin. Not everyone does. Some people have hallucinations. Those little immortal worms, the planarians, when you give them anesthesia (and cut off their heads), regrow a head—of another species. Even bacteria respond to anesthesia.

It's not just drugs that can catch our ion channels by surprise. In 2019, a fifty-four-year-old construction worker collapsed and died even though he was reported to be in perfectly good health. A year later, the *New England Journal of Medicine* published an investigation into the strange case.[19] In the three weeks before he died, the man had eaten one to two large bags of licorice a day, every day. At Massachusetts General Hospital, where doctors spent twenty-four hours trying to save him after his collapse, it became evident that his heart rhythm had been irreversibly destabilized. It turns out that licorice's active ingredient, glycyrrhizin, mimics a process the body uses when it needs to retain sodium and shake off potassium. His potassium channels were gasping for ions, but there weren't any. Without these ions to regulate the sodium–potassium balance of the heart's cells, his heart was incapable of firing regular action potentials. He hadn't been the first to have this experience. Enough similar incidents had piled up by 2012 to occasion a review article entitled "Licorice Abuse," whose authors darkly warned that licorice was "not just a candy." They urged the US Food and Drug Administration to regulate the "substance" and create public health messages around its health hazards.[20] Five years later, the Food and Drug Administration partly obliged, issuing a stark warning about the dangers of licorice in time for Halloween. *Black licorice*, they asked, *trick or treat?*

Quite the tangential mash-up I have assembled here. But I'm trying to paint a picture of all the unexpected ways we can unwittingly affect our electrome. I hope I am making the case for a more holistic understanding of our bioelectric dimensions. Unfortunately, so far there has been a lot of resistance to that. One good example was the reaction to Bates's research.

Institutional silos

The paper shouldn't have been controversial. It was only a review, and a pretty dry one at that. It's not that contentious to point out that bioelectricity seems to play important roles in development, even though the mechanism by which it does so remains unclear. So Bates and her co-author put together an overview of the various mechanisms and theories that have been proposed to account for bioelectrical involvement in fetal development. They sent this draft article to a journal, which distributed it to several other scientists, as is standard practice in peer review in order to publish any paper in a respected scientific journal. The journal editor passed the feedback to Bates, along with a breezy instruction to "address" it. Bates made the mistake of reading the document right before she went to bed.

Some of the responses were scathing in a way that seemed out of all proportion with the paper they were commenting on. No one was talking about methodological flaws or accusing her of fraudulent data. They were simply scathing about the whole field. Phrases like "the mythology of the membrane voltage" suffused the commentary. The fatal blow seemed to be Bates's mention of the bioelectric code.

It wasn't the first time I heard about one of these dismissive, excoriating reviews—Ann Rajnicek (who had worked with Borgens) had told me about a grant application turned down crisply with the single query, "Does anyone really believe this shit anymore?" But as a science journalist, I know that bruising peer review is just part of the game. As I spoke to more researchers, however, I noticed a pattern.[21] People didn't "believe" Laura Hinkle had enacted electrotaxis on cells. Nor did they believe Dany Adams. Or Ai-Sun Tseng. "No one has ever said they don't believe our data," another bioelectricity researcher told me. "They just don't want to hear it." Now here was Bates with another variation on the theme. The common factor seemed to be that critics weren't taking issue with any particulars. Rather, they were issuing sweeping statements, contemptuous dismissals larded with emotion and belief language. While critics often couldn't point to specific problems with the science or processes in the publications, they kept using phrases like "I don't believe it." That's literally what a colleague told Michael Levin at a conference: "I haven't read the papers and I don't need to. I don't believe it."

But what is it they don't believe? That depends on who is doing the disbelieving. Levin is often invited to give talks across a wide range of disciplines, from developmental biology departments to NeurIPs, the biggest AI conference in the world. "Someone always gets pissed off," he said to me. "What they get pissed off about depends on which department I'm in." Contentions that are received as obvious by a neuroscientist are sacrilege for a molecular geneticist. It's no grand conspiracy driving skepticism around bioelectricity's relevance outside the nervous system, though. It's merely education.

Jose Lopez at the National Science Foundation thinks we need new ways to start communicating across these ever-separating disciplines. "We need a new department, and we need polymaths. Not the kind we used to have—that ship has sailed. People like Alexander von Humboldt and Galvani were alive at a time when it was still possible to know everything in science. Now a scientist can spend their entire career on one mutation in one gene that causes one variant of this rare disease." Stephen Badylak agrees that in the field of medicine, many remain in silo.

A compelling new alternative is getting underway right now, exemplified at MIT's Department of Biological Engineering, where you can get a PhD in being a polymath: students in this department are specifically trained to talk across disciplines and focus in on the vocabulary and concepts needed to bridge the chasms between them. They are encouraged to think in a systems biology way about how information is flowing, rather than viewing information as discrete chunks.

"It's so weird that we don't teach this"

Emily Bates got through four years of developmental biology in her undergraduate curriculum at the University of Utah without ever once hearing the term "ion channel." Then she went on to her doctoral work in neuroscience at Harvard, and though ion channels now entered her vocabulary, she never learned about them having any function outside the nervous system. "Of course I knew that they worked for muscle function and also for pancreatic beta cell function," she says, but throughout her many years in undergraduate and graduate education "my perception was that ion channels were studied in neuroscience and that they weren't really studied in other tissues." She was floored when she learned—entirely by happenstance—that channelopathies could cause developmental defects that affect the body's shape and patterning. "I was kind of shocked by that," she says. "It's so weird that we don't teach this."

But she was fascinated by the idea, and so began to work on ion channels in development. But she didn't have any direction. There was no one in her department to guide her. "I thought I was alone, studying this weird thing no one cared about." She didn't even know what terms to put into her literature searches. "I was in kind of a black box."

After the paper was published, she got an email from Michael Levin, who sent her some of his work and introduced her to other people working on similar research. Levin became a hub. Bates started going to conferences and quickly got plugged into a network of other researchers who were looking at her channels too. "Before Michael Levin reached out, I felt like I was just kind of this weird anomaly working on this."

So no wonder her reviewers were so affronted. To be fair, they had probably never been made aware of ion channels either. "It is actually quite difficult to find reviewers with appropriate expertise that agree on anything," Levin said in 2018, during a roundtable meeting held by the editors founding the new journal *Bioelectricity*. "Trying to get reviewers who see the big picture beyond their individual silos has definitely been a challenge." The new journal is part of a movement to make bioelectricity an umbrella discipline in its own right, encompassing the wide range of relevant biological phenomena from developmental biology to AI. The research into bioelectricity, if this project is to succeed, needs to begin to resemble the natural philosophy that prevailed when Galvani made his momentous discoveries. "Since when does nature have departments?" Levin likes to tell people. However, I can't think of a straightforward alternative to dividing science into disciplinary departments.

This separation is part of a modern view of biology that, paradoxically, may be limiting its reach. "Biology today is intensely focused on the molecules of life and particularly on the genes that specify their structures and functions," wrote Franklin Harold in his 2017 book *To Make the World Intelligible*.

But this has limited how we can understand life. One reason bioelectric mechanisms have been so tricky to find—no doubt contributing to bioelectricity's unfair association with quackery—is that the tools to watch these minute and ephemeral processes only became conceivable a few decades ago.

Before that—and even now—looking at live cells has been the exception, not the rule. The majority of scientific discoveries about our biology came from dissecting dead tissues. In a kind of "shoot first, ask questions later" approach, most biological investigations have first killed the cells, then poked around in the goo for the relevant characteristics. And while this has been an excellent way of taxonomizing all the different parts of a cell, dead cells don't do any electric signaling, which has made it virtually impossible to learn about any of the electrical processes that go on in living cells and tissues. Which, in turn, has made it pretty hard to figure out how the electrics affect the other stuff. Looking at cells in this way, writes Paul Davies, is akin to "[trying] to understand how a computer works by studying only the electronics inside it" instead of looking at how these components trade information.[22] People like Galvani and Aldini got lucky that some bioelectric phenomena remain open to investigation for a day or two after death, but in living animals it's been incredibly difficult to observe the flow of electrical currents and change in voltage in real time.

And this is why I am so confident that we are in the bioelectric century. Because the tools that let us look at living cells are advancing at astonishing speed. Just one example is the voltage dye Dany Adams used, which was only developed in the early 2000s. But today, many different labs are using many different approaches to this same kind of technique—making bioelectric parameters observable to the naked eye—and new discoveries are piling up by the day. In 2019, Adam Cohen at Harvard used a fluorescent dye to answer a question that had been bothering him about how cells and tissues transition from their stem identity of a zero voltage to their final electrical identities. Cohen was curious: as an embryo develops, does its voltage glide smoothly from 0 to 70 like a slider, going through all the numbers in between? Or does it snap directly from 0 to 70?

It turned out it was the latter, and that meant that's how entire tissues assume their identity as well: they jump right from stem to bone, not stopping anywhere else along the way. All those cells that are connected by gap junctions transitioned from stem 0 to final state the way ice crystallizes out of water.[23]

There are many new tools in development now that will let us look at living systems in all their electrical complexity in this way, rather than stagnating in the "reductionist fervor" Paul Davies decries in his book.[24]

These will allow us to begin building the full picture of our electrome. Defining the term in 2016, the Dutch biologist Arnold de Loof described "the totality of all ionic currents of any living entity, from the cellular to the organismal level." We'll need to map all our ion channels and gap junctions and figure out how changing cellular voltages can affect cells and tissues. We'll need a visceral nerve atlas to understand how the nervous system controls organ function. I've described much of this work in the book but there's a lot more I didn't have space for. The biophysicist Alexis Pietak has already started developing a tool that can unpick the staggering complexities of how cell voltage leads to cellular identity: a software package called BETSE (the Bioelectric Tissue Simulation Engine) allows people like Michael Levin to simulate how bioelectric signals will propagate in virtual tissue.[25] The dream is that all these tools and the insights they unlock will usher in a future of interfaces that work with—and possibly improve—biology on its own terms.

For the past half century, that honor has gone to machines and engineers who promise a future of all-knowing artificial intelligence, cyborg upgrades to what some have taken to dismissing as our inferior "meat bodies," and a transhumanist deep future in which all biological matter has upgraded to silicon. But recently, the shine has begun to come off AI as we realize just how limited silicon intelligence really is. Existing materials can't even manage hip implants that last longer than ten years—so how are we meant to have a permanent telepathic neural device attached to our brain? The research now underway in bioelectricity suggests that, rather than grasping for silicon and electron replacements to biology, the answers to an upgraded future might lie in biology itself.

Many of the early pioneers of bioelectricity have been redeemed after being initially ignored or derided. This is true not only of

Galvani but also of Harold Saxton Burr, whose predictions about cancer and development have been validated with time, just as Galvani was right about the spark of life. Burr's individual ideas seem to have been broadly right—but in his book published in 1974, he also tied these experiments into a bigger hypothesis. He posited that the day biology investigates forces instead of only studying particles, it will undergo a conceptual leap to rival the importance of splitting the atom for physics.

As we know, that's a double-edged sword.

When we learned about the microbiome, we learned that we could improve it by eating kimchi and lots of greens. The consequence of harnessing the electrome might end up being more complicated.

In fact, the current obsessive focus on hacking our memories and overclocking ourselves into infinite productivity may serve to trap us in the past rather than chart a new future.

Think of it from my perspective. Did tDCS help me overcome an impairment—the constant self-recriminations—or would its regular application constitute an unfair advantage? I'm sure castigating inner voices are hardly a unique feature of my brainscape.

A lot of ink has been spilled asking where you draw the line between medical intervention and cosmetic enhancement. People raise this question all the time about cognitive (and other cosmetic) enhancement, but no one ever seems to come up with a good answer. That's probably because it's actually a question that gets more unsettling the closer you peer at it. Because of course the more people adopt any particular enhancement, the more pressure they will exert on the people around them—and on themselves!—to keep up lest they get left behind, and the more unaugmented normalcy is transformed, by process of inertia, into a deficiency. The blame isn't on any one individual—this is a classic tragedy of the commons.

Particularly in sport, this conversation has been very germane. Discussing the growing use of tDCS in cycling with *Outside* magazine, Thomas Murray, president emeritus of the Hastings Centre bioethics research institute, told the reporter Alex Hutchinson that "once an effective technology gets adopted in a sport, it becomes tyrannical. You have to use it." Hutchinson correctly interpreted this to imply that "if the pros start brain-zapping, don't kid yourself that it won't trickle down to college, high school, and even the weekend warriors."[26] Once you start this game, you can never stop.

So my final exhortation for you, the person who has made it all the way to the end of my book, is this: when you see someone trying to sell you this stuff, ask who will benefit. Why is someone trying to sell it to you? Is it really for you? Beyond the basic "were the trials any good?" skepticism, ask what will happen next. Is this something that will alleviate your suffering? Or will it just kick the can down the road because eventually your new normal will become the new substandard, making way for the next piece of enhancing kit? The answer to that question is very different if the intervention is a treatment for cancer versus a way to be a better hustle goblin consuming more hustle products.

In fact, I would love to take this whole idea of the body as an inferior meat puppet to be augmented with metal and absolutely launch it into the sun. Cybernetics keeps dangling the seductive illusion that we can ascend beyond the grubby world of human biology in our cyborg future—cajoled into correct action and good health (and maximum productivity, of course) by the electrical takeover of a couple of relevant nerve terminals.

The study of the electrome shouldn't serve these masters. Doing the research that led to this book turned my head exactly 180 degrees from this view. Rather than being a collection of inferior meat bodies, biology becomes more astounding the more you learn about it—and fractally complicated too, as the more you learn, the more you realize you don't understand. We are electrical machines whose full dimensions we have not even yet dreamed of.

But as is evident from the MIT program, academia is waking up to the interconnectedness, and different fields are starting to talk to each other more to explore this electric future. That is where we will start to see the next great steps in bioelectricity.

The real excitement of the field hews closer to the excitement around cosmology—a better understanding of our place in the universe and in nature. Even now, the new research is upending some conventional wisdom. I honestly can't wait to see what deeply entrenched the coming decades reveal.

AFTERWORD

Gut feelings: The stomach in your brain

By the time Freda Hall's stomach died, she had already spent six years in unendurable pain. In the early 1990s, in Little Rock, Arkansas, there wasn't much awareness in the medical community that such a thing could even happen. Drugs couldn't help; they never had. Hall's stomach had been giving her trouble all her life—including bouts of vomiting so frequent and severe she couldn't hold down a job. But that morning, an endoscopy clarified the true horror of the situation. Her stomach would not work again. "The doctor was shaking," she says. "Am I going to die?" she recalls asking him. "Yes," he told her helplessly.

A desperate string of useless tests led her around the country. After an otherwise futile investigation at the Mayo Clinic in Jacksonville, Florida, serendipity finally struck when someone there connected enough dots to direct her to the one person in the world who might be able to help her: a gastroenterologist named Tom Abell. He had recently begun clinical trials with adapted Medtronic stimulators—descendants of the early pacemakers—implanting them into the gut in a handful of patients, where they might mitigate the symptoms of a little-known disease called gastroparesis. It was the first treatment that worked on a patient population that, until then, had simply never had options. Those whose symptoms were as severe as Hall's died. Hall was in Abell's office because he had realized something crucial: gastroparesis catastrophically disrupts the natural electrical signals that underpin the gut's ability to digest food. In theory, electrical stimulation should be able to turn them back on. Hall agreed to try the experimental implant.

You'd be forgiven for not knowing there are electrical signals playing their part in your gut health. When you think of the body's electrical activity, the brain or the heart likely have pride of place. Guts, by contrast, conjure churning and acids and chemistry and mechanical motions. But they are also buzzing with their own electrical signals. Until recently these were something of a niche concern. The average person certainly doesn't have much use for a device that corrects the signals that drive the normal rhythm of their digestion—until they go wrong. And this is precisely why you may soon start hearing more about gut electricity. New weight loss drugs are being developed to mess with the pacing and emptying of the stomach. Gastroenterologists are fielding inquiries from these patients about the sort of device Freda Hall first tested to alleviate her symptoms. Some of these doctors don't even know that we have known about the gut's peculiar electrical activity for more than a hundred years.

In 1922, the physician Walter Alvarez predicted a future in which gastroenterologists would come to rely on electrical measures for the routine diagnosis of stomach disorders, "just as the heart specialist, after long experience with the electrocardiograph, can now make shrewd diagnosis of fibrillation, heart block, and so forth."[1] His prognostication was grounded in fact. He had just used the Einthoven galvanometer—famous for demystifying the specific features of the heartbeat—to deduce that there was also electrical activity in the stomach. Indeed, this gut signal had been responsible for some of the interference that tripped up Augustus Waller's first rough electrocardiography readings. Three times a minute, like clockwork, a faint electrical signal shimmered through the indistinct peaks and valleys of the ECG. Eating or fasting; sleeping or awake; that electrical oscillation pulsed steadily on.

As ever, though, the squeaky wheel gets the grease, so the loud, strong, fast signal of the heart got all the attention while the weak and slow gut signal languished. Literally and figuratively, the ECG eclipsed the EGG (electrogastrogram).

Then, in the 1990s, better imaging technology revealed new information about where these signals were coming from: weird cells sitting all along the gut neither muscle nor nerve, known as the interstitial cells of Cajal (ICCs; described as such by Santiago Ramón y Cajal, who initially mistook them for nerves).[2]

Over the past few decades their crucial importance to gut health has become increasingly clear. Just as specialized cells pace the heart, ICCs pace the digestive contractions that travel, wavelike, down the gut. Even their calcium signaling and gap junctions look similar.[3] "Their primary role is just like it is in the heart—to impart a fundamental pattern or rhythm that drives contractions," says Greg O'Grady, a gastrointestinal surgeon at the University of Auckland in New Zealand.

The key difference is that in the heart, every electrical wave results in a commensurate muscle contraction—if it didn't, you'd die. In the stomach, however, while the signal steadily pulses, the muscle contractions join in only as needed, orchestrated by a symphony of neurotransmitters and hormones released in the wake of a meal. When they do, they synchronize precisely to the ICCs' steady electrical tempo.

Well—except when that electricity goes awry.

The worst stomach flu you've ever had

As with most of the disorders you've read about so far in this book, electrical stimulation on the gut started before we knew much about the particulars of its own internal signaling. In the 1970s, a raft of gut-pacing experiments inspired the pacemaker device company Medtronic to approve off-label uses for its devices for bladder and bowel incontinence. Versions are still used today: implanted into the sacral nervous system—a branch of spinal nerves that emerges from its base—next-generation neurostimulators are life-altering for people with bowel incontinence, drastically improving their quality of life.[4] In 80 percent of cases, the devices stop the problem. Unfortunately, things are less straightforward for gastric disorders higher in the gut.

There are many of these, but gastroparesis is by far the worst: it feels as if you constantly have food poisoning, without ever having come down with it. People report severe abdominal bloating and pain, heartburn, constipation, and vomiting. This isn't the kind of vomiting you are thinking of right now, either. You're thinking of the last time you felt nauseated, and you threw up, and as gross as the experience was, it ended, and you felt relief. No such luck for Hall and other people afflicted by the death of their ICCs. "You can throw up all day," she says—and some people do, to the tune of seventeen to twenty-five times a day, or more—but at no point does any of that vomiting offer any relief. "The cascade of weird shit it can cause is just unfathomable to most people," wrote one Reddit contributor. "And you almost can't talk about it without sounding kind of insane."[5]

Some cases of gastroparesis can be traced to diabetes, a result of either autonomic nerve damage or blood sugar spikes that kill the ICCs. But just as often, it can appear without a clear reason, which is how it happened for Freda Hall. In many of those cases, no drugs or dietary changes make a dent. "It's debilitating," says Abell. Even before it shut down completely, Hall's paralyzed gut could neither move food in the correct direction nor—except on rare occasions—extract much nutrition. "I would eat a piece of bread, and it wouldn't come out for nine to eleven days," she says. (In the average healthy person, the ICCs maintain a transit time of about thirty-six hours.) By the time she saw the bread again, it would have been fermented rather than digested. Its form was otherwise largely unchanged.

But she was luckier than most people with the condition. Tom Abell, along with a handful of other scientists, was beginning to suspect that you could fix this situation with artificial bursts of electricity, essentially jump-starting the paralyzed stomach, in cases of severe gastroparesis. Rather than guiding the muscles to contract, as a heart pacemaker does, he increased the frequency of the stimulation so that the device might wake up what few ICCs remained.

When he met Freda Hall, the idea of attacking this problem from an electrical perspective was still so new that before Abell could put a permanent pacemaker into Hall's gut, he first needed to test a temporary version. This device had to be threaded down her throat, a procedure typically done under anesthesia. But Hall had been so toughened up by years of unending misery that she waved away the pain relief.

"I felt it the second he turned it on," she recalls. The ever-present, malevolent awareness of "a dead spot" that had often made it hard for her to take a full breath—it vanished. "Can I eat?" was her first question. Before Abell got the whole word "yes" out of his mouth, Hall says, her husband, Buster, was already out the door to buy her a cheeseburger and fries.

"I got to eat every single day!" she marvels, remembering the months that followed. "I didn't vomit once! I was on top of the world." At the end of that trial period, Abell's surgical team implanted her stimulator into its permanent position. Its near-miraculous ability to help her keep food down inspired an irreverent nickname for her device: "I call him Ralph," she says wryly.

Medtronic preferred the name "Enterra," and twenty years on, this is the device Abell offers his gastroparesis patients when their symptoms are as intractable as Freda Hall's. It is inserted into the anterior stomach under general anesthesia and connected by two wires to a pulse generator sewn into a pouch just under the skin.[6] There, it regulates the electrical chaos that underpins the symptoms of a paralyzed stomach. At least, that's what we think it does. Because gastroparesis is a relatively rare disease, there are too few people like Hall to undertake large trials, which means the effectiveness of the device hasn't been entirely demonstrated—so for the past two decades, the US Food and Drug Administration has authorized it only for humanitarian use.[7] In other words, while they're confident it won't hurt you, they can't say for sure how it works, and they offer zero guarantees that the device will work for anyone.[8]

These and other lingering mysteries have pushed scientists, including Abell and O'Grady, to invent new tools capable of unpicking the intricacies of gut electricity. Electrogastrography readings used to be done using just three electrodes. O'Grady, fed up with this status quo, decided to invent his own improvements. He adapted a tool from neuroscience: using a blanket of 256 EEG electrodes to cover the stomach, he was able to obtain a high-resolution "heat map" of the electrical activity in one of the most inaccessible places in the body.

The first step was to establish, in high definition, a baseline understanding of what "normal" electrical activity in the stomach looked like, courtesy of a batch of healthy volunteers. This group of people had never reported any gut disorders, and they were in the hospital for unrelated abdominal surgery.[9] They consented to being covered with electrodes during their procedure. Armed with this new data,

O'Grady set out to understand what was electrically going on in the guts of people with gastroparesis and other gut disorders.

In about 40 percent of these patients, sure enough, his new high-resolution EGG found evidence of localized electrical damage: in many people with symptoms like Freda Hall's, the ICCs were either malfunctioning or missing outright, a kind of damage that can often accumulate from autoimmune diseases attacking the body's own cells. Their electrical activity, instead of traveling down the gut in regular, synchronized waves, flashed randomly all over their stomachs like a badly pixelated TV signal. This was clear evidence of an electrical dysrhythmia, a problem with a cause that could provide a pathway to being fixed. (For example, a 2023 study found that people with missing ICCs would respond better to stronger energies of electrical stimulation. This is true for Hall; her stimulation strength is so powerful that she has surgery to change the battery every year. It's worth it.)

But O'Grady also found something that was harder to explain: some of the people with gastroparesis had stomach electrical activity that was perfect. The ICCs were pulsing away. What they did have, to a T, were remarkably consistent levels of anxiety and depression. Was there a relationship between this ailment and their uncontrollable gastric symptoms? O'Grady's questions and his findings jibed with a growing realization in the past few years that the relationship between the gut and the brain is not exactly what we thought it was.

The gastric network

"We have this idea that organs are all independent entities," says O'Grady. That's especially misleading when it comes to the gut.

You've probably heard about the "second brain" in your gut,

or the enteric nervous system. This network of 600 million neurons lining our guts has long been investigated for its potential influence on our mental state and decisions as well as its suspected role in a wide range of diseases. It is thought to be a kind of deputy to the brain, a second-in-command independently taking over the daily minutiae of digestion to free up the latter for more-complicated activities like cognition and perception. We didn't have conscious access to gut information because we didn't need it.

But in 2018, Ignacio Rebollo at Sapienza University in Rome discovered that this model of the gut-brain relationship might have missed a spot. And by "a spot" I mean "nodes in the posterior cingulate sulcus, superior parieto-occipital sulcus, dorsal precuneus, retrosplenial cortex, as well as the dorsal and ventral occipital cortex"—areas of the brain where his team found that the slow waves the ICCs produce in the gut were mirrored in perfect synchrony.[10]

Having patterns of activity in the brain that mirrored patterns of activity in the body wouldn't necessarily be a big deal. After all, the brain monitors everything that goes on in the rest of the body, and that produces electrical activity. It shouldn't be surprising that the ICC networks and gut brain axis communicate with each other very actively and can influence each other's behavior, says O'Grady.

But to some, the research implied more than that—it suggested that the famous "second brain" in your stomach had a mirror image: a "second stomach" in your brain; an electric homunculus that was a constant neural representation of the state of the gut, in the brain. People began to call this little stomunculus "the gastric network."[11]

Finding out more about the neural processes that govern this gut-brain connection was a big ask because it involved one of the most inaccessible places in the body for close study. "The gut is like a black box," Maria Inda, a synthetic biologist at Massachusetts Institute of Technology, told *MIT Technology Review*.[12] "The only way we have now is colonoscopy. That's invasive, can't be repeated at short intervals, and disrupts the gut microbiome."

But in 2023, a group led by Sahib Khalsa at the University of Tulsa in Oklahoma found a better way. Volunteers swallowed an edible electronic pill that could be set to vibrate at different frequencies in the gut while the participants were being simultaneously monitored with both EEG and EGG. The mechanical vibrations produced patterns in both—and they mirrored one another. A separate group did the EGG while their volunteers had their brains read by an fMRI, and found that in twelve areas of the brain, the resting state activity was phase-locked to the gut signal, like a puppet master's every small motion being perfectly translated into the movements of their puppet. But was the gut puppeting the gastric network, or was it the other way around?

This question has big implications, especially for illnesses that may have been falsely siloed into having either solely mental or physical origin, says Camilla Nord, a neuroscientist at the University of Cambridge who studies the gut-brain axis and some of the disorders that sit in this liminal space.

Could O'Grady's research imply that some cases of gastroparesis originate from problems farther up the gastric network? And what does that even mean if the brain and the gut each extend so many tendrils into the other? The influence of the body's state on the brain is obvious to anyone who has ever been "hangry," but there are more-subtle consequences. Important and unexplained correlations link certain psychiatric and metabolic disorders. For example, diabetes and depression often co-occur in the same people—someone with diabetes is more likely to have depression than the general population, and vice versa. This concurrence often invites easy, lazy connections—"for example, 'Oh, he's sad so he eats more.' Or, 'Oh, she's obese, so she's sad,'" says Nord. Such connections may seem intuitive, but for Nord, who studies physical symptoms in psychiatric disorders, they are unlikely to tell the full story. "It's more likely that there are also common biological disruptions that could predispose people to both metabolic and mood dysfunction," she says. "And those could be centered around this kind of brain-gut axis interaction."

Another mental illness has clearer physical correlates: a form of post-traumatic stress disorder (PTSD) linked to uncontrollable disgust rather than fear. But unlike fear-linked PTSD, which can be addressed with exposure therapies, disgust-PTSD is impervious.[13] It just won't respond to the usual methods, and Nord suspects this is a consequence of that electrically mediated pacing rhythm of the stomach. The ICCs make sure it stays constant during normal digestion. But recent work shows the rhythm can sometimes change, she says, when you see something disgusting or when you are nauseous. Crucially, this electrical reaction can often be happening below the level of conscious perception. Looking at something you find disgusting changes your stomach's rhythm on an EGG, even if you don't explicitly experience it as nausea.

So Nord did an experiment to see if certain tweaks would make exposure therapy work after all. She gave study participants a drug that keeps stomach rhythms at their normal rate, and then showed them disgusting imagery that had caused them to avert their eyes in an earlier task. Sure enough, after the forced normalization, the photos stopped bothering the volunteers as much, as evidenced by their eyes, which no longer shied away.[14] "By changing the state of someone's stomach," Nord reported in her book *The Balanced Brain*, it finally became possible to habituate them to disgusting imagery, offering a potential route to better mental health via the stomach.

A more pressing problem has arisen since 2021: the clinic where Tom Abell works has seen an influx of patients whose symptoms stem from an unusual new source.

One weird trick to lose weight

You have probably seen the relentless blitz of advertising for injectable weight-loss drugs, with opaque names like Wegovy, Ozempic, and Mounjaro. It all started with Ozempic, developed in the mid-2010s for people with type 2 diabetes, in whom it lowers and controls blood sugar. After increasing evidence that people on Ozempic also reliably lost weight, the drug was soon conscripted into off-label use by people who wanted to budge stubborn pounds for non-medical reasons, including actors and influencers whose public victories over their body fat launched the products into the public imagination. When the drug's manufacturer, Novo Nordisk, realized the potential size of this market, it developed a specific formulation of the drug for weight loss. Wegovy—the same ingredient in a larger dose—entered the scene with promises of fast and dramatic weight loss. Competing manufacturers soon released similar drugs called Mounjaro and Zepbound. No one can keep these drugs in stock, so further entrants to the market are inevitable.

All these varied names boil down to the same mechanism: a way to impersonate a naturally occurring hormone, GLP-1, secreted by the gut in response to eating. Stuffing it into the body's "Am I still hungry?" receptors comes with the obvious result that you feel immediately and extremely full, even after you've eaten very little food. Often, the effect can tip into extremes: across official medical reports and anonymous Reddit posts alike, people have reported feeling nauseated instead of satiated; they report diarrhea or constipation; and they experience excruciating pain and "delayed gastric emptying," a deceptively anodyne medical phrase that describes the sensation of undigested food putrefying in your stomach. Many people have justified these experiences as necessary evils because of what they see as the drugs' salutary effects, which can include not just weight loss but—according to a Novo Nordisk–funded clinical trial—significantly better cardiovascular health.[15]

Some try to wait out the symptoms, but for others, they only get worse. "By the time they come to see us, these people are really miserable," says Abell. They don't want to just stop taking the drug and forfeit its perceived benefits. So instead, many are now beginning to inquire about the Enterra device to take the misery out of their gut.

It's a weird situation for Abell because these people seem to have given themselves the worst stomach disorder he can imagine. Its nastiest effects sure look a lot like the classic presentation of gastroparesis. GLP-1 drugs are administered to cause pseudogastroparesis.[16] But how long can you keep your body in pseudogastroparesis before it just becomes . . . gastroparesis?

That's the question raised by a lawsuit filed against Novo Nordisk and Eli Lilly in 2023,[17] which alleged that the drugs caused gastroparesis in several patients, who detailed a familiar hideous bouquet of experiences: one had been vomiting days-old food; another had to have a stomach bypass surgery to help relieve the gastroparesis.[18] In their motion to dismiss the complaint, the drug makers said the symptoms are known and documented as side effects on the drug's approved label.[19]

The consequences of the little stomach in your brain may go beyond medicine and veer into philosophy, as they may also offer new insights into why some of our emotions manifest so physically:

stress as an IBS (irritable bowel syndrome) trigger; depression and anxiety as tummy sensitivity. "Think about the electrical communication between the gut and the brain in the same way you think about the electrical communication between the heart and the brain," says Nord. "These can really affect the experience of your emotions." Dread can turn your guts to liquid. Disgust makes your stomach churn.

An army of ingestible sensor pills and eavesdropping equipment is being deployed to understand the complicated chicken-and-egg relationship between the physical and mental dimensions of having an emotion. But beyond simply *affecting* our emotions, the connections in the gastric network may even have shaped the entire structure and evolution of emotions themselves. They may even yield clues to how our shared subjective experiences of emotions formed in the first place.

One thing is clear: we have only started to scratch the surface of the mysteries the gastric network may reveal, but simultaneous brain and gut monitoring will fast-track the new insights. And the rapid spread of drugs like Wegovy, Ozempic, and Mounjaro may help provide answers to some of these ongoing mysteries around the relationship between gut electricity and the gastric network. One obvious step would be to monitor their effects on an EGG, which no one has done. "Someone really should," says Abell. Around the world, gastroparesis affects about 0.9 percent of the general population. In the United States that number is highest in the world, double the worldwide average at 1.7 percent[20]—and those numbers were collected before seemingly half the country asked their doctor about Ozempic. The number will rise. Freda Hall wishes it wouldn't. "Fifteen pounds is not worth it," she says. She is now seventy-two and spry. She credits Abell with saving her life. She can't imagine how anyone can volunteer for what nature conscripted her into.

The work of researchers like Abell and O'Grady is untangling ever more of the complexities of the gut electrome, revealing the electrical properties of cells beyond the standard nerve and muscle. Nord and Khalsa are delving into the more complicated electrical interactions that connect the mind and the body. The gut is not the only place bioelectric discoveries are changing previous assumptions. The body generates and makes use of electric fields in many ways, and we are starting to be able to mimic these—and no longer just to make fake action potentials in the nervous system. The migration of tens of thousands of cells can now be directed electrically, which will accelerate wound healing.[21] Electrical guidance systems are even being deployed to help plants grow better.[22]

As all these different fields break silos, they will spur more such previously unlikely collaborations and crossovers. Our electric future will never stop yielding new surprises.

ACKNOWLEDGMENTS

There's a joke that first-time book authors thank every person they have ever met and I will be no exception. First, thanks to Simon Thorogood for taking a chance on this! To Mollie Weisenfeld and Georgia Frances King for executing the most enormous trust fall. To Carrie Plitt for seeing the outlines of the book long before I ever did.

I cannot get over the generosity of the scientists and researchers who answered my emails and phone and Zoom queries. They spent hours with me through Covid lockdowns and electoral chaos to explain wildly difficult concepts, talk me through controversial history, and help me keep my focus. Alphabetically: Dany Adams patiently read draft after draft, and sent diagrams and markups, and ironed out my misunderstandings. Debra Bohnert, thank you so much for our wonderful conversation—I hope the story does the man justice. Robert Campenot fielded my first terrified questions with grace and good humor. Edward Farmer, thank you for reading many drafts of chapters that were cut! I promise you will see them again. Flavio Frohlich, our conversations unlocked an entire chapter. Franklin Harold, thank you for giving me the most wonderful quote that had to be cut out of the final version of the book (I will never not laugh about the vibrating probe at Wood's Hole, and I'm not sorry). Andrew Jackson, thank you for making me start thinking seriously about ions. Nancy Kopell, thank you for an unforgettable conversation. Michael Levin: thank you for four years of tirelessly answering questions, sending papers, reading drafts, sending more papers, reading more drafts, et cetera ad infinitum. Jianming Li, thank you for making me understand the hair-raisingly difficult mechanics of die-back and for the many hours on the phone. Kip Ludwig, thank you for sending more papers than a single human could read in one lifetime, and for "fast email answers" that went on for multiple pages. Marco Piccolino, thank you for helping me figure out where to start, and then making sure I ended up at the right end—thank you in particular for sending me your amazing book. Ann Rajnicek, thank you for giving me the oral history of Lionel Jaffe's lab, but most of all for telling me about Richard Borgens. Ken Robinson spent ages on the phone with me to explain how the spinal stimulator did (and did not) work. Nigel Wallbridge, for all the time spent explaining the bioelectric CANBUS system. Harold Zakon and Min Zhao, thank you for explaining, respectively, ion-channel motifs and galvanotaxis.

Phone calls can't do everything. I also trudged through some difficult reading to try to understand the complicated history of how neuroscience integrated electricity. Some books offered the keys that unlocked these papers and historical documents. Among these, three stand out for their clarity and explanatory power: Matthew Cobb's book *The Idea of the Brain* was an invaluable resource. So were Robert Campenot's *Animal Electricity*, and Frances Ashcroft's *Spark of Life*. For anyone in whom my book sparked a curiosity about the history of science of the brain and nervous system, I urge them to drop everything and read these books.

To the people who read my early drafts, I am indebted to you forever: Richard Panek, thank you so much for your inestimable help with my first, unreadable attempts at "history of science."

ACKNOWLEDGMENTS

Lowri Daniels and Michele Kogon, you are rock stars—thank you for ensuring my facts were actually facts. David Robson, Clare Wilson, Richard Fisher, thanks for patiently listening to my spittle-flecked terror. Darryl Rambo, Lorri Lofvers, and Joyce Wong, for being a steadfast lifeboat of calm in turbulent years. Sumit Paul-Choudhury, Hal Hodson, and Will Heaven, thank you for the life-saving powers of Voltron, and for enduring my endless yammering without exiting the chat (well, thanks to two of you anyway). Sarita Bhatt: for understanding the subtext. Cristina Calotta—I can't thank you enough for the endless stream of requests for GDPR articles I couldn't access from their US websites or a library during Covid. Soren, Cassie, Erin, Mike—you're the constellations in the firmament. Ann, you're the North Star.

Dad, you got me curious about all the things there are to know in the world—no good deed goes unpunished, so now you have to read this whole book. Mom, you made sure I had a skeptical head on my shoulders while I fell down every rabbit hole. And Nick and Daisy and Charlie: thank you for enduring the late nights, the caffeine jitters, the endless illegible notes about the neural code scrawled on random pieces of paper (and sometimes on important mail), the mood rollercoasters. Thank you for your encouragement, your grace, your patience, and your love.

NOTES

Introduction

1. Condliffe, Jamie. "Glaxo and Verily Join Forces to Treat Disease By Hacking Your Nervous System," *MIT Technology Review*, 1 August 2016. <https://www.technologyreview.com/2016/08/01/158574/glaxo-and-verily-join-forces-to-treat-disease-by-hacking-your-nervous-system>
2. Hutchinson, Alex. "For the Golden State Warriors, Brain Zapping Could Provide an Edge," *The New Yorker*, 15 June 2016. <https://www.newyorker.com/tech/annals-of-technology/for-the-golden-state-warriors-brain-zapping-could-provide-an-edge>
3. Reardon, Sarah. "'Brain doping' may improve athletes' performance." *Nature* 531 (2016), pp. 283–4
4. Farmer, Edward E., et al. "Wound- and Mechanostimulated Electrical Signals Control Hormone Responses." *New Phytologist*, vol. 227, no. 4, (2020), pp. 1037–1050
5. Blackiston, Douglas J., and Michael Levin. "Ectopic eyes outside the head in Xenopus tadpoles provide sensory data for light-mediated learning." *Journal of Experimental Biology* 216 (2013), pp. 1031–40; Durant, Fallon, et al. "Long-Term, Stochastic Editing of Regenerative Anatomy via Targeting Endogenous Bioelectric Gradients." *Biophysical Journal*, vol. 112, no. 10 (2017), pp. 2231–43

Part 1: Bioelectricity in the Beginning

Chapter 1: Artificial vs Animal

1. Pancaldi, Giuliano. *Volta: Science and Culture in the Age of Enlightenment.* Princeton, NJ: Princeton University Press, 2005, p. 111

2   Galvani, Luigi. *Commentary on the Effects of Electricity on Muscular Motion.* Trans. Margaret Glover Foley. Norwalk, CN: Burndy Library, 1953, p. 79
3   Pancaldi, *Volta*, p. 54; and Morus, Iwan Rhys. *Frankenstein's Children: Electricity, Exhibition, and Experiment in Early-Nineteenth-Century London.* Princeton, NJ: Princeton University Press, 1998, p. 232
4   Needham, Dorothy. *Machina Carnis: The Biochemistry of Muscular Contraction in its Historical Development.* Cambridge: Cambridge University Press, 1971, pp. 1–26
5   Needham, *Machina Carnis*, p. 7
6   Kinneir, David. "A New Essay on the Nerves, and the Doctrine of the Animal Spirits Rationally Considered." London, 1738, pp. 21 and 66–7 <https://archive.org/details/b30525068/page/n5/mode/2up>
7   O'Reilly, Michael Francis, and James J. Walsh. *Makers of Electricity.* New York: Fordham University Press, 1909, p. 81
8   Cohen, I. Bernard. *Benjamin Franklin's Science.* Cambridge, MA: Harvard University Press, 1990, p. 42
9   Finger, Stanley, and Marco Piccolino. *The Shocking History of Electric Fishes.* Oxford: Oxford University Press, 2011, pp. 282–5
10 Bresadola, Marco, and Marco Piccolino. *Shocking Frogs: Galvani, Volta, and the Electric Origins of Neuroscience.* Oxford: Oxford University Press, 2013, p. 27
11 Bergin, William. "Aloisio (Luigi) Galvani (1737–1798) and Some Other Catholic Electricians." In: Sir Bertram Windle (ed.), *Twelve Catholic Men of Science.* London: Catholic Truth Society, 1912, pp. 69–87
12 Bresadola & Piccolino, *Shocking Frogs*, p. 27
13 O'Reilly & Walsh, *Makers of Electricity*, p. 152; and Bergin, "Aloisio (Luigi) Galvani," p. 75
14 Cavazza, Marta. "Laura Bassi and Giuseppe Veratti: an electric couple during the Enlightenment." *Institut d'Estudis Catalans*, vol. 5, no. 1 (2009), pp. 115–24 (pp. 119–21)
15 Messbarger, R. M. *The Lady Anatomist: The Life and Work of Anna Morandi Manzolini.* Chicago: University of Chicago Press, 2010, p. 157
16 Frize, Monique. *Laura Bassi and Science in 18th-Century Europe.* Berlin/Heidelberg: Springer, 2013; see also Messbarger, *The Lady Anatomist*, pp. 171–3

17 Focaccia, Miriam, and Raffaella Simili. "Luigi Galvani, Physician, Surgeon, Physicist: From Animal Electricity to Electro-Physiology." In: Harry Whitaker, C. U. M. Smith and Stanley Finger (eds), *Brain, Mind and Medicine: Essays in Eighteenth-Century Neuroscience.* Boston: Springer, 2007, pp. 145–58 (p. 154)
18 Bresadola & Piccolino, *Shocking Frogs*, p. 76
19 Bresadola & Piccolino, *Shocking Frogs*, p. 89
20 Bresadola & Piccolino, *Shocking Frogs*, p. 122
21 O'Reilly & Walsh, *Makers of Electricity*, p. 133
22 See Bernardi, W. "The controversy on animal electricity in eighteenth-century Italy. Galvani, Volta and others." In: F. Bevilacqua and L. Fregonese (eds), *Nuova Voltiana: Studies on Volta and His Times Vol. 1.* Milan: Hoepli, 2000, pp. 101–12 (p. 102). A translation is available at <http://www.edumed.org.br/cursos/neurociencia/controversy-bernardi.pdf; and Bresadola & Piccolino, *Shocking Frogs*, p. 143, among others
23 Pancaldi, *Volta*, pp. 14–15
24 Pancaldi, *Volta*, p. 20
25 Pancaldi, *Volta*, p. 31
26 Pancaldi, *Volta*, p. 91
27 Pancaldi, *Volta*, p. 111
28 Pancaldi, *Volta*, p. 111
29 Bresadola & Piccolino, *Shocking Frogs*, p. 152
30 Bresadola & Piccolino, *Shocking Frogs*, pp. 143–4
31 Bernardi, "The controversy," pp. 104–5
32 Material about the French commissions from Blondel, Christine. "Animal Electricity in Paris: From Initial Support, to Its Discredit and Eventual Rehabilitation." In: Marco Bresadola and Giuliano Pancaldi (eds), *Luigi Galvani International Workshop*, 1998, pp. 187–204
33 Blondel, "Animal Electricity," p. 189
34 Volta, Alessandro. "Memoria seconda sull'elettricita animale" (14 May 1792). Quoted in: Pera, Marcello. *The Ambiguous Frog.* Trans. Jonathan Mandelbaum. Princeton, NJ: Princeton University Press, 1992, p. 106
35 Unless otherwise referenced, the quotes from the rash of scientific papers in this section have been taken from Bresadola & Piccolino, *Shocking Frogs* and Pera, *The Ambiguous Frog*

36 Ashcroft, Frances. *The Spark of Life*. London: Penguin, 2013, p. 24
37 Blondel, "Animal Electricity," p. 190
38 Bernardi, "The controversy," p. 107 (fn. 26)
39 Robert Campenot provides a clear and straightforward description of this experiment. Campenot, Robert. *Animal Electricity*. Cambridge, MA: Harvard University Press, 2016, p. 40
40 Bernardi, "The controversy," p. 103
41 Bernardi, "The controversy," p. 107

Chapter 2: Spectacular pseudoscience

1 Aldini, Giovanni. *Essai théorique et expérimental sur le galvanisme, avec une série d'expériences. Faites en présence des commissaires de l'Institut National de France, et en divers amphithéâtres Anatomiques de Londres*. Paris: Fournier Fils, 1804. Available via the Smithsonian Libraries archive at <https://library.si.edu/digital-library/book/essaitheyorique00aldi>

2 Some sources suggest Queen Charlotte and her son, the Prince of Wales, attended but it may have been the younger prince, Augustus Frederick, who Aldini later dedicated a book to. It seems clear that there was at least one royal present.

3 Tarlow, Sarah, and Emma Battell Lowman. *Harnessing the Power of the Criminal Corpse*. London: Palgrave Macmillan, 2018, pp. 87–114

4 McDonald, Helen. "Galvanising George Foster, 1803," The University of Melbourne Archives and Special Collections. <https://library.unimelb.edu.au/asc/whats-on/exhibitions/dark-imaginings/gothicresearch/galvanising-george-foster,-1803>

5 Morus, Iwan Rhys. *Frankenstein's Children: Electricity, Exhibition, and Experiment in Early-Nineteenth-Century London*. Princeton, NJ: Princeton University Press, 1998, p. 128

6 Sleigh, Charlotte. "Life, Death and Galvanism." *Studies in History and Philosophy of Science Part C: Studies in History and Philosophy of Biological and Biomedical Sciences*, vol. 29, no. 2 (1998), pp. 219–48 (p. 223)

7 There are many accounts of this experiment—mine is drawn mainly from Morus, Iwan Rhys. *Shocking Bodies: Life, Death & Electricity in Victorian England*. Stroud: The History Press, 2011,

pp. 34–7. Other sources are Aldini's personal account and the *Newgate Calendar*, 22 January 1803, p. 3
8   Sleigh, "Life, Death and Galvanism," p. 224
9   Parent, André. "Giovanni Aldini: From Animal Electricity to Human Brain Stimulation," *Canadian Journal of Neurological Sciences / Journal Canadien des Sciences Neurologiques*, vol. 31, no. 4 (2004), pp. 576–84 (p. 578)
10 Blondel, Christine. "Animal Electricity in Paris: From Initial Support, to Its Discredit and Eventual Rehabilitation." In: Marco Bresadola and Giuliano Pancaldi (eds), *Luigi Galvani International Workshop*, 1998, pp. 187–204 (pp. 194–5)
11 Aldini, "Essai Théorique," p. vi
12 Aldini's most detailed account of such a treatment concerns Luigi Lanzarini.
13 Carpue, Joseph. "An Introduction to Electricity and Galvanism; with Cases, Shewing Their Effects in the Cure of Diseases." London: A. Phillips, 1803, p. 86 <https://wellcomecollection.org/works/bzaj37cs/items?canvas=100>
14 Blondel, "Animal Electricity," p. 197
15 Aldini, John [sic]. "General Views on the Application of Galvanism to Medical Purposes, Principally in Cases of Suspended Animation." London: Royal Society, 1819, p. 37. When publishing abroad, Aldini had the habit of changing his first name. In the UK he anglicized to John, and in France he became Jean.
16 Parent, "Giovanni Aldini," p. 581
17 Vassalli-Eandi said in August 1802 that Aldini "has been obliged to acknowledge that he had not been able to get any contractions from the heart using the electro motor of Volta."
18 Aldini, "Essai Théorique," p. 195
19 Giulio, C. "Report presented to the Class of the Exact Sciences of the Academy of Turin, 15th August 1802, in Regard to the Galvanic Experiments Made by C. Vassali-Eandi, Giulio and Rossi on the 10th and 14th of the same Month, on the Head and Trunk of three Men a short Time after their Decapitation." *The Philosophical Magazine*, vol. 15, no. 57 (1803), pp. 39–41
20 Morus, Iwan. "The Victorians Bequeathed Us Their Idea of an Electric Future." *Aeon*, 8 August 2016
21 Aldini, "Essai Théorique," p. 143–4

22 This section draws heavily from: Bertucci, Paola. "Therapeutic Attractions: Early Applications of Electricity to the Art of Healing." In: Harry Whitaker, C. U. M. Smith, and Stanley Finger (eds), *Brain, Mind, and Medicine: Essays in Eighteenth-Century Neuroscience*. Boston: Springer, 2007, pp. 271–83; Pera, Marcello, *The Ambiguous Frog*. Trans. Jonathan Mandelbaum. Princeton, NJ: Princeton University Press, 1992; and several unbeatable details from Iwan Rhys Morus's *Frankenstein's Children*
23 Pera, *The Ambiguous Frog*, pp. 18–25
24 Pera, *The Ambiguous Frog*, p. 22
25 Ashcroft, Frances. *The Spark of Life*. London: Penguin, 2013, pp. 290–1
26 Bertucci, "Therapeutic Attractions," p. 281
27 Calculated on 23 May 2022 using the CPI Inflation Calculator. <https://www.officialdata.org/uk-inflation>
28 Bertucci, "Therapeutic Attractions," p. 281
29 Shepherd, Francis John. "Medical Quacks and Quackeries," *Popular Science Monthly*, vol. 23 (June 1883), p. 152
30 Morus, *Shocking Bodies*, p. 35
31 Ochs, Sidney. *A History of Nerve Functions: From Animal Spirits to Molecular Mechanisms*. Cambridge: Cambridge University Press, 2004, p. 117
32 Miller, William Snow. "Elisha Perkins and His Metallic Tractors." *Yale Journal of Biology and Medicine*, vol. 8, no. 1 (1935), pp. 41–57 (p. 44)
33 Lord Byron. "English Bards and Scotch Reviewers." Quoted in: Miller, "Elisha Perkins," p. 52
34 Finger, Stanley, Marco Piccolino, and Frank W. Stahnisch. "Alexander von Humboldt: Galvanism, Animal Electricity, and Self-Experimentation Part 2: The Electric Eel, Animal Electricity, and Later Years." *Journal of the History of the Neurosciences*, vol. 22, no. 4 (2013), pp. 327–52 (p. 343)
35 Finger, Stanley, and Marco Piccolino. *The Shocking History of Electric Fishes*. Oxford: Oxford University Press, 2011, p. 11
36 Finger et al., "Alexander von Humboldt," p. 343
37 Otis, Laura. *Müller's Lab*. Oxford: Oxford University Press, 2007 p. 11; see also Finger et al., "Alexander von Humboldt," p. 345
38 A picture can be seen at "Nobili's large astatic galvanometer," Museo Galileo Virtual Museum <https://catalogue.museogalileo.it/object/NobilisLargeAstaticGalvanometer.html>

39 Verkhratsky, Alexei, and Parpura, Vladimir. "History of Electrophysiology and the Patch Clamp." In: Marzia Martina and Stefano Taverna (eds), *Methods in Molecular Biology*. New York: Humana Press, 2014, pp. 1–19 (p. 7). However, much of the detail about Nobili and Matteucci's experiments comes from Otis's *Müller's Lab*.
40 Cobb, Matthew. *The Idea of the Brain: A History*. London: Profile Books, 2020, p. 71
41 Finger et al., "Alexander von Humboldt," p. 347 and Otis, p. 90
42 Emil du Bois-Reymond in an 1849 letter to fellow experimental physiologist Carl Ludwig, reproduced on p. 347 of: Finger et al., "Alexander von Humboldt."
43 Finger & Piccolino, *The Shocking History of Electric Fishes*, p. 369
44 Bresadola, Marco, and Marco Piccolino. *Shocking Frogs: Galvani, Volta, and the Electric Origins of Neuroscience*. Oxford: Oxford University Press, 2013, p. 21
45 Finkelstein, Gabriel. "Emil du Bois-Reymond vs Ludimar Hermann." *Comptes rendus biologies*, vol. 329, 5-6 (2006), pp. 340-7 doi:10.1016/j.crvi.2006.03.005

Part 2: Bioelectricity and the Electrome

1 Bresadola, Marco, and Marco Piccolino. *Shocking Frogs: Galvani, Volta, and the Electric Origins of Neuroscience*. Oxford: Oxford University Press, 2013, p. 13

Chapter 3: The electrome and the bioelectric code

1 The first mention of the word "electrome" can be found in an obscure 2016 paper written by the Belgian biologist Arnold de Loof ("The cell's self-generated 'electrome': The biophysical essence of the immaterial dimension of Life?," *Communicative & Integrative Biology*, vol. 9,5, e1197446). This definition did not break into wider circulation. Even before its publication, however, other bioelectricity researchers, including Michael Levin and Min Zhao, had begun to use the word. Zhao, in particular, has reviewed a few manuscripts using that term "without [consistent] definition, and clarification. It is an evolving understanding." The purpose of this book is to pin the word down like a butterfly behind glass.

2   Valenstein, Elliot. *The War of the Soups and the Sparks: The Discovery of Neurotransmitters and the Dispute over how Nerves Communicate*. New York: Columbia University Press, 2005, pp. 121–34
3   James, Frank. "Davy, Faraday, and Italian Science." Report presented at the IX National Conference of "History and Foundations of Chemistry" (Modena, 25–27 October 2001), pp. 149–58 <https://media.accademiaxl.it/memorie/S5-VXXV-P1-2-2001/James149-158.pdf> Accessed 22 February 2021
4   Faraday, Michael. *Experimental Researches in Electricity—Volume 1* [1832]. London: Richard and John Edward Taylor, 1849. Available at <https://www.gutenberg.org/files/14986/14986-h/14986-h.htm>
5   Ringer, Sydney, and E. A. Morshead. "The Influence on the Afferent Nerves of the Frog's Leg from the Local Application of the Chlorides, Bromides, and Iodides of Potassium, Ammonium, and Sodium." *Journal of Anatomy and Physiology* 12 (October 1877), pp. 58–72
6   Campenot, Robert, *Animal Electricity*. Cambridge, MA: Harvard University Press, 2016, p. 114
7   McCormick, David A. "Membrane Potential and Action Potential." In: Larry Squire et al. (eds), *Fundamental Neuroscience*. Oxford: Academic Press, 2013, pp. 93–116 (p. 93)
8   Hodgkin, Alan, and Andrew F. Huxley. "A quantitative description of membrane current and its application to conduction and excitation in nerve." *The Journal of Physiology*, vol. 117, no. 4 (1952), pp. 500–44
9   Bresadola, Marco, and Marco Piccolino. *Shocking Frogs: Galvani, Volta, and the Electric Origins of Neuroscience*. Oxford: Oxford University Press, 2013, p. 294
10 Ramachandran, Vilayanur S. "The Astonishing Francis Crick." Francis Crick memorial lecture delivered at the Centre for the Philosophical Foundations of Science in New Delhi, India, 17 October 2004. <http://cbc.ucsd.edu/The_Astonishing_Francis_Crick.htm>
11 Schuetze, Stephen. "The Discovery of the Action Potential." *Trends in Neurosciences* 6 (1983), pp. 164–8. See also Lombard, Jonathan, "Once upon a time the cell membranes: 175 years of cell boundary research." *Biology Direct*, vol. 9, no. 32, pp. 1–35; and Finger, Stanley, and Marco Piccolino. *The Shocking History of Electric Fishes*. Oxford: Oxford University Press, 2011, p. 402
12 Campenot, *Animal Electricity*, pp. 210–11

13 Agnew, William, et al. "Purification of the Tetrodotoxin-Binding Component Associated with the Voltage-Sensitive Sodium Channel from Electrophorus Electricus Electroplax Membranes." *Proceedings of the National Academy of Sciences*, vol. 75, no. 6 (1978), pp. 2606–10

14 Noda, Masaharu, et al. "Expression of Functional Sodium Channels from Cloned CDNA." *Nature*, vol. 322, no. 6082 (1986), pp. 826–8.

15 Brenowitz, Stephan, et al. "Ion Channels: History, Diversity, and Impact." *Cold Spring Harbor Protocols* 7 (2017), loc. pdb.top092288 <http://cshprotocols.cshlp.org/content/2017/7/pdb.top092288.long#sec-3>

16 McCormick, "Membrane Potential and Action Potential," p. 103

17 Ashcroft, Frances. *The Spark of Life*. London: Penguin, 2013, p. 69

18 McCormick, David A. "Membrane Potential and Action Potential." In: John H. Byrne, and James L. Roberts (eds), *From Molecules to Networks: An Introduction to Cellular and Molecular Neuroscience*, 2nd edition. Amsterdam/Boston: Academic Press, 2009, pp. 133–58 (p. 151)

19 Ashcroft, *The Spark of Life*, p. 49 and pp. 87–9

20 Barhanin, Jacques, et al. "New scorpion toxins with a very high affinity for Na+ channels. Biochemical characterization and use for the purification of Na+ channels." *Journal de Physiologie*, vol. 79, no. 4 (1984), pp. 304–8

21 Kullmann, Dimitri M. "The Neuronal Channelopathies." *Brain*, vol. 125, no. 6 (2002), pp. 1177–95

22 Fozzard, Harry. "Cardiac Sodium and Calcium Channels: A History of Excitatory Currents." *Cardiovascular Research*, vol. 55, no. 1 (2002), pp. 1–8

23 Sherman, Harry G., et al. "Mechanistic insight into heterogeneity of trans-plasma membrane electron transport in cancer cell types." *Biochimica et Biophysica Acta—Bioenergetics*, 1860/8 (2019), pp. 628–39

24 Lund, Elmer. *Bioelectric Fields and Growth*. Austin: University of Texas Press, 1947

25 Prindle A, Liu J, Asally M, Ly S, Garcia-Ojalvo J, Süel GM. "Ion channels enable electrical communication in bacterial communities." *Nature*. (2015) Nov 5;527(7576):59-63. doi: 10.1038/nature15709. Epub 2015 Oct 21. PMID: 26503040; PMCID: PMC4890463

26 Brand, Alexandra et al. "Hyphal Orientation of Candida albicans

Is Regulated by a Calcium-Dependent Mechanism." *Current Biology*, 17, (2007), pp. 347–352
27 Davies, Paul. *The Demon in the Machine*. London: Allen Lane, 2019, p. 110
28 Anderson, Paul A., and Robert M. Greenberg. "Phylogeny of ion channels: clues to structure and function." *Comparative Biochemistry and Physiology Part B: Biochemistry and Molecular Biology*, vol. 129, no. 1 (2001), pp. 17–28. doi: 10.1016/s1096-4959(01)00376-1
29 Liebeskind, B. J., D. M. Hillis, and H. H. Zakon. "Convergence of ion channel genome content in early animal evolution." *Proceedings of the National Academies of Science* 112 (2015), E846–E851

Part 3: Bioelectricity in the Brain and Body

Chapter 4: Electrifying the heart

1 Besterman, Edwin, and Creese, Richard. "Waller—pioneer of electrocardiography." *British Heart Journal*, vol. 42, no. 1 (1979), pp. 61–4 (p. 63)
2 Acierno, Louis. "Augustus Desire Waller." *Clinical Cardiology*, vol. 23, no. 4 (2000), pp. 307–9 (p. 308)
3 Harrington, Kat. "Heavy browed savants unbend." Royal Society blogs, 14 July 2016. Retrieved from the Internet Archive 21 September 2021 <https://web.archive.org/web/20191024235429/http://blogs.royalsociety.org/history-of-science/2016/07/04/heavy-browed/>
4 Waller, Augustus D. "A Demonstration on Man of Electromotive Changes accompanying the Heart's Beat." *The Journal of Physiology*, vol. 8 (1887), pp. 229–34
5 Campenot, Robert. *Animal Electricity*. Cambridge, MA: Harvard University Press, 2016, p. 269
6 Burchell, Howard. "A Centennial Note on Waller and the First Human Electrocardiogram." *The American Journal of Cardiology*, vol. 59, no. 9 (1987), pp. 979–83 (p. 979)
7 AlGhatrif, Majd, and Joseph Lindsay. "A Brief Review: History to Understand Fundamentals of Electrocardiography." *Journal of Community Hospital Internal Medicine Perspectives*, vol. 2 no. 1 (2012), loc. 14383

8   Ashcroft, Frances. *The Spark of Life*. London: Penguin, 2013, p. 146
9   Campenot, *Animal Electricity*, pp. 272–4
10 Aquilina, Oscar. "A brief history of cardiac pacing." *Images in Paediatric Cardiology*, vol. 8, no. 2 (April 2006), pp. 17–81 (Fig. 16)
11 Rowbottom, Margaret, and Charles Susskind. *Electricity and Medicine: History of Their Interaction*. London: Macmillan, 1984, p. 248
12 Rowbottom & Susskind, *Electricity and Medicine*, p. 249
13 Rowbottom & Susskind, *Electricity and Medicine*, p. 249
14 Emery, Gene. "Nuclear pacemaker still energized after 34 years," Reuters, 19 December 2007 <https://www.reuters.com/article/us-heart-pacemaker-idUSN1960427320071219>
15 Norman, J. C. et al. "Implantable nuclear-powered cardiac pacemakers." *New England Journal of Medicine*, vol. 283, no. 22 (1970), pp. 1203–6. doi: 10.1056/NEJM197011262832206
16 Roy, O. Z., and R. W. Wehnert. "Keeping the heart alive with a biological battery." *Electronics*, vol. 39, no. 6 (1966), pp. 105–7. Also see: <https://link.springer.com/article/10.1007/BF02629834>
17 Greatbatch, Wilson. *The Making of the Pacemaker: Celebrating a Lifesaving Invention*. Amherst: Prometheus Books, 2000, p. 23
18 Tashiro, Hiroyuki, et al. "Direct Neural Interface." In: Marko B. Popovic (ed.), *Biomechatronics*. Oxford: Academic Press, 2019, pp. 139–74
19 Greatbatch, *The Making of the Pacemaker*, p. 23

Chapter 5: Artificial memories and sensory implants

1   Hamzelou, Jessica. "$100 million project to make intelligence-boosting brain implant," *New Scientist*, 20 October 2016 <https://www.newscientist.com/article/2109868-100-million-project-to-make-intelligence-boosting-brain-implant/>
2   McKelvey, Cynthia. "The Neuroscientist Who's Building a Better Memory for Humans," *Wired*, 1 December 2016 <https://www.wired.com/2016/12/neuroscientist-whos-building-better-memory-humans/>
3   Johnson, Bryan. "The Urgency of Cognitive Improvement," *Medium*, 14 June 2017 <https://medium.com/future-literacy/the-urgency-of-cognitive-improvement-72f5043ca1fc>
4   Campenot, Robert, *Animal Electricity*. Cambridge, MA: Harvard University Press, 2016, pp. 110–11

5   Finger, Stanley. *Minds Behind the Brain*. Oxford: Oxford University Press, 2005, pp 243–7. See also Ashcroft, Frances. *The Spark of Life*. London: Penguin, 2013, ch. 3
6   Garson, Justin. "The Birth of Information in the Brain: Edgar Adrian and the Vacuum Tube." *Science in Context*, vol. 28, no. 1 (2015), pp. 31–52 (pp. 40–2)
7   Finger, *Minds*, p. 249
8   Finger, *Minds*, p. 250
9   Finger, *Minds*, p. 250
10 Garson, "The Birth," p. 46
11 Finger, *Minds*, p. 250
12 Adrian, E. D. *The Physical Background of Perception*. Quoted in Cobb, Matthew. *The Idea of the Brain: A History*. London: Profile Books, 2020, p. 186
13 Borck, Cornelius. "Recording the Brain at Work: The Visible, the Readable, and the Invisible in Electroencephalography." *Journal of the History of the Neurosciences* 17 (2008), pp. 367–79 (p. 371)
14 Millett, David. "Hans Berger: From Psychic Energy to the EEG." *Perspectives in Biology and Medicine*, vol. 44, no. 4 (2001), pp. 522–42 (p. 523)
15 Ginzberg, quoted in Millet, "Hans Berger," p. 524
16 Millet, "Hans Berger," p. 537
17 Cobb, *The Idea of the Brain*, p. 170
18 Millet, "Hans Berger," p. 539
19 Borck, "Recording," p. 369
20 Borck, "Recording," p. 368
21 Borck, Cornelius, and Ann M. Hentschel. *Brainwaves: A Cultural History of Electroencephalography*. London: Routledge, 2018, p. 110
22 Borck & Hentschel, *Brainwaves*, p. 109
23 Borck & Hentschel, *Brainwaves*, p. 115
24 Collura, Thomas. "History and Evolution of Electroencephalographic Instruments and Techniques." *Journal of Clinical Neurophysiology*, vol. 10, no. 4 (1993), pp. 476–504 (p. 498)
25 Marsh, Allison. "Meet the Roomba's Ancestor: The Cybernetic Tortoise," IEEE Spectrum, 28 February 2020 <https://spectrum.ieee.org/meet-roombas-ancestor-cybernetic-tortoise>
26 Cobb, *The Idea of the Brain*, p. 190
27 Hodgkin, Alan. "Edgar Douglas Adrian, Baron Adrian of

NOTES

Cambridge. 30 November 1889–4 August 1977." *Biographical Memoirs of Fellows of the Royal Society* 25 (1979), pp. 1–73 (p. 19)

28 Tatu, Laurent. "Edgar Adrian (1889–1977) and Shell Shock Electrotherapy: A Forgotten History?." *European Neurology*, vol. 79, nos. 1–2 (2018), pp. 106–7

29 Underwood, Emil. "A Sense of Self." *Science*, vol. 372, no. 6547 (2021), pp. 1142–5 (pp. 1142–3)

30 Olds, James. "Pleasure Centers in the Brain." *Scientific American*, vol. 195 (1956), pp. 105–17; Olds, James. "Self-Stimulation of the Brain." *Science* 127 (1958), pp. 315–24

31 Moan, Charles, and Robert G. Heath. "Septal Stimulation for the Initiation of Heterosexual Behavior in a Homosexual Male." In: Wolpe, Joseph, and Leo J. Reyna (eds), *Behavior Therapy in Psychiatric Practice*. New York: Pergamon Press, 1976, pp. 109–16

32 Giordano, James (ed.). *Neurotechnology*. Boca Raton: CRC Press, 2012, p. 151

33 Frank, Lone. "Maverick or monster? The controversial pioneer of brain zapping," *New Scientist*, 27 March 2018 <https://www.newscientist.com/article/mg23731710-700-maverick-or-monster-the-controversial-pioneer-of-brain-zapping/>

34 Blackwell, Barry. "José Manuel Rodriguez Delgado." *Neuropsychopharmacology*, vol. 37, no. 13 (2012), pp. 2883–4.

35 The photo has been widely reproduced, but can be found in Marzullo, Timothy. "The Missing Manuscript of Dr. José Delgado's Radio Controlled Bulls." *JUNE*, vol. 15, no. 2 (Spring 2017), pp. 29–35

36 Osmundsen, John. "Matador with a radio stops wired bull: modified behavior in animals subject of brain study." *New York Times*, 17 May 1965

37 Horgan, John. "Tribute to José Delgado, Legendary and Slightly Scary Pioneer of Mind Control." *Scientific American*, 25 September 2017

38 Gardner, John. "A History of Deep Brain Stimulation: Technological Innovation and the Role of Clinical Assessment Tools." *Social Studies of Science*, vol. 43, no. 5 (2013), pp. 707–28 (p. 710)

39 Schwalb, Jason M., and Clement Hamani. "The History and Future of Deep Brain Stimulation." *Neurotherapeutics*, vol. 5, no. 1 (2008), pp. 3–13

40 Gardner, "A History," p. 719
41 Lozano, A. M., N. Lipsman, H. Bergman, et al. "Deep brain stimulation: Current challenges and future directions." *Nature Reviews Neurology* 15 (2019), pp. 148–60 <https://www.nature.com/articles/s41582-018-0128-2>
42 Nuttin, Bart, et al. "Electrical Stimulation in Anterior Limbs of Internal Capsules in Patients with Obsessive-Compulsive Disorder." *The Lancet*, vol. 354, no. 9189 (1999), p. 1526
43 Ridgway, Andy. "Deep brain stimulation: A wonder treatment pushed too far?," *New Scientist*, 21 October 2015 <https://www.newscientist.com/article/mg22830440-500-deep-brain-stimulation-a-wonder-treatment-pushed-too-far/>
44 Sturm, V., et al. "DBS in the basolateral amygdala improves symptoms of autism and related self-injurious behavior: a case report and hypothesis on the pathogenesis of the disorder." *Frontiers in Neuroscience*, vol. 6, no. 341 (2013), doi: 10.3389/fnhum.2012.00341
45 Formolo, D. A., et al. "Deep Brain Stimulation for Obesity: A Review and Future Directions." *Frontiers in Neuroscience*, vol. 13, no. 323 (2019), doi: 10.3389/fnins.2019.00323; Wu, H., et al. "Deep-brain stimulation for anorexia nervosa." *World Neurosurgery* 80 (2013), doi: 10.1016/j.wneu.2012.06.039
46 Baguley, David, et al. "Tinnitus." *The Lancet*, vol. 382, no. 9904 (2013), pp. 1600–7; Luigjes, J., van den Brink, W., Feenstra, M., et al. "Deep brain stimulation in addiction: a review of potential brain targets." *Molecular Psychiatry* 17 (2012), pp. 572–83 <https://doi.org/10.1038/mp.2011.114>; Fuss, J., et al. "Deep brain stimulation to reduce sexual drive." *Journal of Psychiatry and Neuroscience*, vol. 40, no. 6 (2015) pp. 429–31
47 Satellite meeting of Society for Neuroscience, San Diego, 2018. Mayberg also spoke about it at the Brain & Behaviour Research Foundation: "Deep Brain Stimulation for Treatment-Resistant Depression: A Progress Report," Brain & Behaviour Research Foundation YouTube channel, 16 October 2019 <https://www.youtube.com/watch?v=X86wBj1tjiA>
48 Mayberg, Helen, et al. "Deep Brain Stimulation for Treatment-Resistant Depression." *Neuron*, vol. 45, no. 5 (2005), pp. 651–60
49 Dobbs, David. "Why Deep-Brain Stimulation for Depression

Didn't Pass Clinical Trials," *The Atlantic*, 17 April 2018 <https://www.theatlantic.com/science/archive/2018/04/zapping-peoples-brains-didnt-cure-their-depression-until-it-did/558032/>

50 "BROADEN Trial of DBS for Treatment-Resistant Depression No Better than Sham," The Neurocritic blog, 10 October 2017 <https://neurocritic.blogspot.com/2017/10/broaden-trial-of-dbs-for-treatment.html>

51 "The Remote Control Brain," *Invisibilia*, NPR, first broadcast 29 March 2019 <https://www.npr.org/2019/03/28/707639854/the-remote-control-brain>

52 Cyron, Donatus. "Mental Side Effects of Deep Brain Stimulation (DBS) for Movement Disorders: The Futility of Denial." *Frontiers in Integrative Neuroscience* 10 (2016), pp. 1–4 <https://www.frontiersin.org/articles/10.3389/fnint.2016.00017/full>

53 Mantione, Mariska, et al. "A Case of Musical Preference for Johnny Cash Following Deep Brain Stimulation of the Nucleus Accumbens." *Frontiers in Behavioral Neuroscience*, vol. 8, no. 152 (2014), doi: 10.3389/fnbeh.2014.00152

54 Florin, Esther, et al. "Subthalamic Stimulation Modulates Self-Estimation of Patients with Parkinson's Disease and Induces Risk-Seeking Behaviour." *Brain*, vol. 136, no. 11 (2013), pp. 3271–81.

55 Shen, Helen H., "Can Deep Brain Stimulation Find Success beyond Parkinson's Disease?" *Proceedings of the National Academy of Sciences*, vol. 116, no. 11 (2019), pp. 4764–6

56 Müller, Eli J., and Peter A. Robinson. "Quantitative Theory of Deep Brain Stimulation of the Subthalamic Nucleus for the Suppression of Pathological Rhythms in Parkinson's Disease," ed. by Saad Jbabdi, *PLOS Computational Biology*, vol. 14, no. 5 (2018), e1006217. See also Kisely, Steve, et al. "A Systematic Review and Meta-Analysis of Deep Brain Stimulation for Depression." *Depression and Anxiety*, vol. 35, no. 5 (2018), pp. 468–80

57 Crick, Francis. *The Astonishing Hypothesis: The Scientific Search for the Soul*. New York: Scribner; London: Maxwell Macmillan International, 1994, p. 10, see also pp. 182-4

58 Crick, *The Astonishing Hypothesis*, p. 3. For more on consciousness, a wonderful resource is Chapter 15 of Matthew Cobb's *The Idea of the Brain*

59 Gerstner, Wulfram, et al. "Neural Codes: Firing Rates and Beyond." *Proceedings of the National Academy of Sciences*, vol. 94, no. 24 (1997), pp. 12740–1 <https://www.pnas.org/doi/epdf/10.1073/pnas.94.24.12740>
60 See Buzsöki, Gyárgy. *Rhythms of the Brain*. New York: Oxford University Press, 2011
61 Kellis, Spencer, et al. "Decoding Spoken Words Using Local Field Potentials Recorded from the Cortical Surface." *Journal of Neural Engineering*, vol. 7, no. 5 (2010), 056007
62 Martin, Richard. "Mind Control," *Wired*, 1 March 2005 <https://www.wired.com/2005/03/brain-3/>
63 Martin, "Mind Control," 2005
64 Bouton, Chad. "Reconnecting a paralyzed man's brain to his body through technology," TEDx Talks YouTube channel, 25 November 2014 <https://www.youtube.com/watch?v=BPI7XWPSbS4>
65 Bouton, C., Shaikhouni, A., Annetta, N., et al. "Restoring cortical control of functional movement in a human with quadriplegia." *Nature* 533 (2016), pp. 247–50 <https://doi.org/10.1038/nature17435>
66 Geddes, Linda. "First paralysed person to be 'reanimated' offers neuroscience insights," *Nature*, 13 April 2016 <https://doi.org/10.1038/nature.2016.19749>
67 Geddes, Linda. "Pioneering brain implant restores paralysed man's sense of touch," *Nature*, 13 October 2016 <https://doi.org/10.1038/nature.2016.20804>
68 Flesher, S. N., et al. "Intracortical microstimulation of human somatosensory cortex." *Science Translational Medicine*. vol. 8, no. 361 (2016), doi: 10.1126/scitranslmed.aaf8083
69 Berger, T. W., et al. "A cortical neural prosthesis for restoring and enhancing memory." *Journal of Neural Engineering*, vol. 8, no. 4 (2011), doi: 10.1088/1741-2560/8/4/046017
70 Frank, Loren. "How to Make an Implant That Improves the Brain," *MIT Technology Review*, 9 May 2013 <https://www.technologyreview.com/2013/05/09/178498/how-to-make-a-cognitive-neuroprosthetic/>
71 Hampson, Robert E., et al. "Facilitation and Restoration of Cognitive Function in Primate Prefrontal Cortex by a Neuroprosthesis That Utilizes Minicolumn-Specific Neural Firing." *Journal of Neural Engineering*, vol. 9, no. 5 (2012), 056012

72 Strickland, Eliza. "DARPA Project Starts Building Human Memory Prosthetics," IEEE Spectrum, 27 August 2014 <https://spectrum.ieee.org/darpa-project-starts-building-human-memory-prosthetics>
73 McKelvey, "The Neuroscientist," 2016
74 Ganzer, Patrick, et al. "Restoring the Sense of Touch Using a Sensorimotor Demultiplexing Neural Interface." *Cell*, vol. 181, no. 4 (2020) pp. 763–73
75 "Reconnecting the Brain After Paralysis Using Machine Learning," *Medium*, 21 September 2020 <https://medium.com/mathworks/reconnecting-the-brain-after-paralysis-using-machine-learning-1a134c622c5d>
76 Bryan, Carla, and Ivan Rios (eds). *Brain–machine Interfaces: Uses and Developments*. New York: Novinka, 2018
77 Chad Bouton is working on the solution to the "take home" problem. Bouton, Chad. "Brain Implants and Wearables Let Paralyzed People Move Again," IEEE Spectrum, 26 January 2021 <https://spectrum.ieee.org/brain-implants-and-wearables-let-paralyzed-people-move-again>
78 Engber, Daniel. "The Neurologist Who Hacked His Brain—And Almost Lost His Mind." *Wired*, 26 January 2016
79 Jun, James J., et al. "Fully Integrated Silicon Probes for High-Density Recording of Neural Activity." *Nature*, vol. 551, no. 7679 (2017), pp. 232–6
80 Strickland, Eliza. "4 Steps to Turn 'Neural Dust' Into a Medical Reality," IEEE Spectrum, 21 October 2016 <https://spectrum.ieee.org/4-steps-to-turn-neural-dust-into-a-medical-reality>
81 Lee, Jihun, et al. "Neural Recording and Stimulation Using Wireless Networks of Microimplants." *Nature Electronics*, vol. 4, no. 8 (2021), pp. 604–14
82 "Brain chips will become 'more common than pacemakers,' says investor, as startup raises $10m," The Stack, 19 May 2021 <https://thestack.technology/blackrock-neurotech-brain-machine-interfaces-peter-thiel/>
83 Drew, Liam. "Elon Musk's Neuralink brain chip: what scientists think of first human trial." *Nature*, 2 February 2024. <https://www.nature.com/articles/d41586-024-00304-4>
84 Ghose, Carrie. "Ohio State researcher says Battelle brain-computer interface for paralysis could save $7B in annual home-care costs,"

Columbus Business First, 10 October 2019 <https://www.bizjournals.com/columbus/news/2019/10/10/ohio-state-researcher-saysbattelle-brain-computer.html>

85 Regalado, Antonio. "Thought Experiment," *MIT Technology Review*, 17 June 2014 <https://www.technologyreview.com/2014/06/17/172276/the-thought-experiment/>

Chapter 6: The healing spark

1 Bowen, Chuck. "Nerve Repair Innovation Gives Man Hope," Spinal Cord Injury Information Pages, 4 July 2007 <https://www.sci-info-pages.com/news/2007/07/nerve-repair-innovation-gives-man-hope/>
2 Wallack, Todd. "Sense of urgency for spinal device," *Boston Globe*. 18 September 2007 <http://archive.boston.com/business/globe/articles/2007/09/18/sense_of_urgency_for_spinal_device/>
3 Per Debra Bohnert, Richard Borgens's lab director from 1986 to 2019, in a telephone interview with the author.
4 Jaffe, L. F., and M.-m. Poo. "Neurites grow faster towards the cathode than the anode in a steady field." *Journal of Experimental Zoology* 209 (1979), pp. 115–28
5 Ingvar, Sven. "Reaction of cells to the galvanic current in tissue cultures." *Experimental Biology and Medicine*, vol. 17, no. 8 (1920)
6 Bishop, Chris. "The Briks of Denton and Dallas TX," Garage Hangover, 18 October 2007 <https://garagehangover.com/briks-denton-dallas/>
7 Pithoud, Kelsey. "Ex-rocker turns to research," *The Purdue Exponent*, 17 September 2003 <https://web.archive.org/web/20151216205707/https://www.purdueexponent.org/campus/article_73f34375-9059-5273-b6a8-8d9577c74b5d.html>
8 Bishop, "The Briks," 2007
9 Comment by Johnny Young on Bishop, "The Briks," 2007. 25 January 2019 at 11.33 a.m.
10 Kolsti, Nancy. "This is ... Spinal Research," The North Texan Online, Fall 2001 <https://northtexan.unt.edu/archives/f01/spinal.htm>
11 Hinkle, Laura, et al. "The direction of growth of differentiating neurones and myoblasts from frog embryos in an applied electric field." *The Journal of Physiology*, 314 (1981), pp. 121–35

12 McCaig, Colin. "Epithelial Physiology, Ovarian Follicles, Nerve Growth Cones, Vibrating Probes, Wound Healing, and Cluster Headache: Staggering Steps on a Route Map to Bioelectricity." *Bioelectricity*, vol. 2, no. 4 (2020), pp. 411–17 (p. 412)
13 Borgens, Richard, et al. "Bioelectricity and Regeneration." *BioScience*, vol. 29, no. 8 (1979), pp. 468–74
14 Borgens, Richard, et al. "Large and persistent electrical currents enter the transected lamprey spinal cord." *Proceedings of the National Academy of Sciences*, vol. 77, no. 2 (1980), pp. 1209–13
15 Borgens, Richard B., Andrew R. Blight, and M. E. McGinnis. "Behavioral Recovery Induced by Applied Electric Fields After Spinal Cord Hemisection in Guinea Pig." *Science*, vol. 238, no. 4825 (1987), pp. 366–9
16 Kleitman, Naomi. "Under one roof: the Miami Project to Cure Paralysis model for spinal cord injury research." *Neuroscientist*, vol. 7, no. 3 (2001), pp. 192–201
17 Borgens, Richard B., et al. "Effects of Applied Electric Fields on Clinical Cases of Complete Paraplegia in Dogs." *Restorative Neurology and Neuroscience*, vol. 5, nos. 5–6 (1993), pp. 305–22
18 "Electrical stimulation helps dogs with spinal injuries," *Purdue News*, 21 July 1993 <https://www.purdue.edu/uns/html3month/1990-95/930721.Borgens.dogstudy.html>
19 Orr, Richard. "Research On Dogs' Spinal Cord Injuries May Lead To Help For Humans," *Chicago Tribune*, 20 November 1995 <https://www.chicagotribune.com/news/ct-xpm-1995-11-20-9511200137-story.html>
20 "Purdue/IU partnership in paralysis research," Purdue News Service, 28 July 1999 <https://www.purdue.edu/uns/html4ever/1999/990730.Borgens.institute.html>
21 "Human Trial for Spinal Injury Treatment Launched by Purdue, IU," Purdue News Service, December 2000 <https://www.purdue.edu/uns/html4ever/001120.Borgens.SpinalTrial.html>
22 Callahan, Rick. "Two universities launch clinical trial for paralysis patients," *Middletown Press*, 12 December 2000 <https://www.middletownpress.com/news/article/Two-universities-launch-clinical-trial-for-11940807.php>
23 This quote is taken from an edition of Purdue School of Veterinary Medicine's self-published newsletter seen by the author: "Tales

from the Vet Clinic: Yukon overcomes his chilling ordeal!" *Synapses*, Fall 2020
24 "Device to Aid Paralysis Victims to Get Test," *Los Angeles Times*, 13 December 2000
25 Bowen, C. "Nerve Repair Innovation Gives Man Hope," *Indianapolis Star*, 4 July 2007
26 Ravn, Karen. "In spinal research, pets lead the way," *Los Angeles Times*, 9 April 2007 <https://www.latimes.com/archives/la-xpm-2007-apr-09-he-labside9-story.html>
27 "Implanted device offers new sensation," *The Engineer*, 11 January 2005 <https://www.theengineer.co.uk/implanted-device-offers-new-sensation/>
28 "Cyberkinetics to acquire Andara Life Science for $4.5M," *Boston Business Journal*, 13 February 2006 <https://www.bizjournals.com/boston/blog/mass-high-tech/2006/02/cyberkinetics-to-acquire-andara-life-science.html>
29 Cyberkinetics press release, 28 September 2006 <https://www.purdue.edu/uns/html3month/2006/060928CyberkineticsAward.pdf>
30 Robinson, Kenneth, and Peter Cormie. "Electric Field Effects on Human Spinal Injury: Is There a Basis in the In Vitro Studies?" *Developmental Neurobiology*, vol. 68, no. 2 (2008), pp. 274–80
31 Wallack, "Sense of urgency," 2007
32 Shapiro, Scott. "A Review of Oscillating Field Stimulation to Treat Human Spinal Cord Injury." *World Neurosurgery*, vol. 81, nos. 5–6 (2014), pp. 830–5
33 Bowman, Lee. "Study on dogs yields hope in human paralysis treatment," *Seattle Post-Intelligencer*, 3 August 2004
34 Li, Jianming. "Oscillating Field Electrical Stimulator (OFS) for Regeneration of the Spinal Cord," 2017 entry to the Create the Future Design Contest <https://contest.techbriefs.com/2017/entries/medical/8251>
35 Li, Jianming. "Weak Direct Current (DC) Electric Fields as a Therapy for Spinal Cord Injuries: Review and Advancement of the Oscillating Field Stimulator (OFS)." *Neurosurgical Review*, vol. 42, no. 4 (2019), pp. 825–34
36 Willyard, Cassandra. "How a Revolutionary Technique Got People with Spinal-Cord Injuries Back on Their Feet." *Nature*, vol. 572, no. 7767 (2019), pp. 20–5

37 Even chemical and physical factors like contact inhibition release and population pressure.
38 McCaig, Colin D., et al. "Controlling Cell Behavior Electrically: Current Views and Future Potential." *Physiological Reviews*, vol. 85, no. 3 (2005), pp. 943–78
39 "Direct-current (DC) electric fields are present in all developing and regenerating animal tissues, yet their existence and potential impact on tissue repair and development are largely ignored," they wrote in "Controlling Cell Behavior Electrically."
40 Reid, Brian, et al. "Wound Healing in Rat Cornea: The Role of Electric Currents." *The FASEB Journal*, vol. 19, no. 3 (2005), pp. 379–86
41 Hagins, W. A., et al. "Dark Current and Photocurrent in Retinal Rods." *Biophysical Journal*, vol. 10, no. 5 (1970), pp. 380–412
42 Song, Bing, et al. "Electrical Cues Regulate the Orientation and Frequency of Cell Division and the Rate of Wound Healing in Vivo." *Proceedings of the National Academy of Sciences*, vol. 99, no. 21 (2002), pp. 13577–82
43 Leppik, Liudmila, et al. "Electrical Stimulation in Bone Tissue Engineering Treatments." *European Journal of Trauma and Emergency Surgery*, vol. 46, no. 2 (2020), pp. 231–44
44 Zhao, Min, et al. "Electrical Signals Control Wound Healing through Phosphatidylinositol-3-OH Kinase-γ and PTEN." *Nature*, vol. 442, no. 7101 (2006), pp. 457–60.
45 See National Institutes for Health, "A Clinical Trial of Dermacorder for Detecting Malignant Skin Lesions," 17 November 2009 <https://clinicaltrials.gov/ct2/show/NCT01014819>
46 Nuccitelli, R., et al. "The electric field near human skin wounds declines with age and provides a noninvasive indicator of wound healing." *Wound Repair and Regeneration*, vol. 19, no. 5 (2011), pp. 645–55
47 Stephens, Tim. "Bioelectronic device achieves unprecedented control of cell membrane voltage," UC Santa Cruz News Center, 24 September 2020 <https://news.ucsc.edu/2020/09/bioelectronics.html>
48 Ershad, F., A. Thukral., J. Yue, et al. "Ultra-conformal drawn-on-skin electronics for multifunctional motion artifact-free sensing and point-of-care treatment." *Nature Communications*, vol. 11, no. 3823 (2020), doi: https://doi.org/10.1038/s41467-020-17619-1

Part 4: Bioelectricity in Birth and Death

Chapter 7: In the beginning

1. Levin, Michael. "What Bodies Think About: Bioelectric Computation Beyond the Nervous System as Inspiration for New Machine Learning Platforms." The Thirty-second Annual Conference on Neural Information Processing Systems (NIPS). Palais des Congrès de Montréal, Montréal, Canada. 4 December 2018, slide 49 <https://media.neurips.cc/Conferences/NIPS2018/Slides/Levin_bioelectric_computation.pdf >; see also Pullar, Christine E. (ed.). *The Physiology of Bioelectricity in Development, Tissue Regeneration and Cancer*. Boca Raton: CRC Press, 2011, p. 69
2. Sámpogna, Gianluca, et al. "Regenerative Medicine: Historical Roots and Potential Strategies in Modern Medicine." *Journal of Microscopy and Ultrastructure*, vol. 3, no. 3 (2015), pp. 101–7 (p. 101)
3. Power, Carl, and John E. J. Rasko. "The stem cell revolution isn't what you think it is." *New Scientist*, 29 September 2021 <https://www.newscientist.com/article/mg25133542-600-the-stem-cell-revolution-isnt-what-you-think-it-is>
4. Burr, Harold Saxton, et al. "A Vacuum Tube Micro-Voltmeter for the Measurement of Bio-Electric Phenomena." *The Yale Journal of Biology and Medicine*, vol. 9, no. 1 (1936), pp. 65–76. It is pictured on the journal's website alongside the article: <https://www.ncbi.nlm.nih.gov/pmc/articles/PMC2601500/figure/F1/>
5. Burr, Harold Saxton. *Blueprint for Immortality: The Electric Patterns of Life*. Essex: Neville Spearman Publishers, 1972, p. 48
6. Burr, Harold Saxton, L. K. Musselman, Dorothy Barton, and Naomi B. Kelly. "Bio-Electric Correlates of Human Ovulation." *The Yale Journal of Biology and Medicine*, vol. 10, no. 2 (1937), pp. 155–60
7. Burr, Harold Saxton, R. T. Hill, and E. Allen. "Detection of Ovulation in the Intact Rabbit." *Proceedings of the Society for Experimental Biology and Medicine*, vol. 33, no. 1 (1935), pp. 109–11
8. Burr, *Blueprint*, p. 50
9. Burr, *Blueprint*, p. 51
10. Langman, Louis, and H. S. Burr. "Electrometric Timing of Human Ovulation." *American Journal of Obstetrics and Gynecology*, vol. 44, no. 2 (1942), pp. 223–9

NOTES

11 "Medicine: Yale Proof," *Time*, 11 October 1937 <http://content.time.com/time/subscriber/article/0,33009,770949-1,00.html>

12 There is a diagram on p. 156 of Burr et al., "Bio-Electric Correlates." <https://www.ncbi.nlm.nih.gov/pmc/articles/PMC2601785/?page=2>

13 Altmann, Margaret. "Interrelations of the Sex Cycle and the Behavior of the Sow." *Journal of Comparative Psychology*, vol. 31, no. 3 (1941), pp. 481–98

14 "Dr. John Rock (1890–1984)," PBS American Experience <https://www.pbs.org/wgbh/americanexperience/features/pill-dr-john-rock-1890-1984/>

15 Snodgrass, James, et al. "The Validity of 'Ovulation Potentials.'" *American Journal of Physiology—Legacy Content*, vol. 140, no. 3 (1943), pp. 394–415

16 Su, Hsiu-Wei, et al. "Detection of Ovulation, a Review of Currently Available Methods." *Bioengineering & Translational Medicine*, vol. 2, no. 3 (2017), pp. 238–46

17 Herzberg, M., et al. "The Cyclic Variation of Sodium Chloride Content in the Mucus of the Cervix Uteri." *Fertility and Sterility*, vol. 15, no. 6 (1964), pp. 684–94

18 Burr, Harold Saxton, and L. K. Musselman. "Bio-Electric Phenomena Associated with Menstruation." *The Yale Journal of Biology and Medicine*, vol. 9, no. 2 (1936), pp. 155–8

19 Tosti, Elisabetta. "Electrical Events during Gamete Maturation and Fertilization in Animals and Humans." *Human Reproduction Update*, vol. 10, no. 1 (2004), pp. 53–65

20 Van Blerkom, J. "Domains of High-Polarized and Low-Polarized Mitochondria May Occur in Mouse and Human Oocytes and Early Embryos." *Human Reproduction*, vol. 17, no. 2 (2002), pp. 393–406

21 Trebichalská, Zuzana, and Zuzana Holubcová. "Perfect Date—the Review of Current Research into Molecular Bases of Mammalian Fertilization." *Journal of Assisted Reproduction and Genetics*, vol. 37, no. 2 (2020), pp. 243–56

22 Stein, Paula, et al. "Modulators of Calcium Signalling at Fertilization." *Open Biology*, vol. 10, no. 7 (2020), loc. 200118

23 Campbell, Keith H., et al. "Sheep cloned by nuclear transfer from a cultured cell line." *Nature*, vol. 380, article 6569 (1996), pp. 64–6 (p. 64)

24 Zimmer, Carl. "Growing Left, Growing Right," *The New York Times*,

3 June 2013 <https://www.nytimes.com/2013/06/04/science/growing-left-growing-right-how-a-body-breaks-symmetry.html>
25   Some have problems with breathing normally and fertility.
26   See Nuccitelli, Richard, *Ionic Currents In Development*. New York: International Society of Developmental Biologists, 1986
27   Tosti, E., R. Boni, and A. Gallo. "Ion currents in embryo development." *Birth Defects Research Part C* 108 (2016), pp. 6–18, doi: 10.1002/bdrc.21125
28   Adams, Dany S., and Michael Levin. "General Principles for Measuring Resting Membrane Potential and Ion Concentration Using Fluorescent Bioelectricity Reporters." *Cold Spring Harbor Protocols*, 2012/4 (2012)
29   Cone, Clarence, and Charlotte M. Cone. "Induction of Mitosis in Mature Neurons in Central Nervous System by Sustained Depolarization." *Science*, vol. 192, no. 4235 (1976), pp. 155–8
30   Knight, Kalimah Redd, and Patrick Collins, "The Face of a Frog: Time-lapse Video Reveals Never-Before-Seen Bioelectric Pattern." Tufts University press release, 18 July 2011 <https://now.tufts.edu/2011/07/18/face-frog-time-lapse-video-reveals-never-seen-bioelectric-pattern>
31   Vandenberg, Laura N., et al. "V-ATPase-Dependent Ectodermal Voltage and Ph Regionalization Are Required for Craniofacial Morphogenesis." *Developmental Dynamics*, vol. 240, no. 8 (2011), pp. 1889–904
32   Adams, Dany Spencer, et al. "Bioelectric Signalling via Potassium Channels: A Mechanism for Craniofacial Dysmorphogenesis in KCNJ2-Associated Andersen-Tawil Syndrome: K + -Channels in Craniofacial Development." *The Journal of Physiology*, vol. 594, no. 12 (2016), pp. 3245–70
33   Moody, William J., et al. "Development of ion channels in early embryos." *Journal of Neurobiology* 22 (1991) pp. 674–84
34   Rovner, Sophie. "Recipes for Limb Renewal," *Chemical & Engineering News*, 2 August 2010 <https://pubsapp.acs.org/cen/science/88/8831sci1.html>
35   Pai, Vaibhav P., et al. "Transmembrane Voltage Potential Controls Embryonic Eye Patterning in Xenopus Laevis." *Development*, vol. 139, no. 2 (2012), pp. 313–23
36   Malinowski, Paul T., et al. "Mechanics dictate where and how freshwater planarians fission." *PNAS*, vol. 114, no. 41 (2017), pp.

10888–93 <www.pnas.org/cgi/doi/10.1073/pnas.1700762114>
37 Hall, Danielle. "Brittle Star Splits," Smithsonian Ocean, January 2020 <https://ocean.si.edu/ocean-life/invertebrates/brittle-star-splits>
38 Levin, Michael. "Reading and Writing the Morphogenetic Code: Foundational White Paper of the Allen Discovery Center at Tufts University," p. 2 <https://allencenter.tufts.edu/wp-content/uploads/Whitepaper.pdf>
39 Kolata, Gina. "Surgery on Fetuses Reveals They Heal Without Scars," *The New York Times*, 16 August 1988 <https://www.nytimes.com/1988/08/16/science/surgery-on-fetuses-reveals-they-heal-without-scars.html>
40 Barbuzano, Javier. "Understanding How the Intestine Replaces and Repairs Itself," *Harvard Gazette*, 14 July 2017 <https://news.harvard.edu/gazette/story/2017/07/understanding-how-the-intestine-replaces-and-repairs-itself/>
41 Vanable, Joseph. "A history of bioelectricity in development and and regeneration." In: Charles E. Dinsmore (ed.), *A History of Regeneration Research*. New York: Cambridge University Press, 1991, pp. 151–78 (p. 163)
42 Sisken, Betty. "Enhancement of Nerve Regeneration by Selected Electromagnetic Signals." In: Marko Markov (ed.), *Dosimetry in Bioelectromagnetics*, Boca Raton: CRC Press, 2017, pp. 383–98
43 Tseng A.-S., et al. "Induction of Vertebrate Regeneration by a Transient Sodium Current." *Journal of Neuroscience*, vol. 30, no. 39 (2010), pp. 13192–13200
44 Tseng, Ai-sun, and Michael Levin. "Cracking the bioelectric code: Probing endogenous ionic controls of pattern formation." *Communicative & Integrative Biology*, vol. 6, no. 1 (2013): e22595
45 Eskova, Anastasia, et al. "Gain-of-Function Mutations of Mau / DrAqp3a Influence Zebrafish Pigment Pattern Formation through the Tissue Environment." *Development* 144 (2017), doi:10.1242/dev.143495
46 Dlouhy, Brian J., et al. "Autograft-Derived Spinal Cord Mass Following Olfactory Mucosal Cell Transplantation in a Spinal Cord Injury Patient: Case Report." *Journal of Neurosurgery: Spine*, vol. 21, no. 4 (2014), pp. 618–22
47 Jabr, Ferris. "In the Flesh: The Embedded Dangers of Untested Stem Cell Cosmetics," *Scientific American*, 17 December 2012 <https://www.scientificamerican.com/article/stem-cell-cosmetics/>

48 Aldhous, Peter. "An Experiment That Blinded Three Women Unearths the Murky World of Stem Cell Clinics," BuzzFeed News, 21 March 2017 <https://www.buzzfeednews.com/article/peter-aldhous/stem-cell-tragedy-in-florida>
49 Coghlan, Andy. "How 'stem cell' clinics became a Wild West for dodgy treatments," *New Scientist*, 17 January 2018 <https://www.newscientist.com/article/mg23731610-100-how-stem-cell-clinics-became-a-wild-west-for-dodgy-treatments/>
50 Feng J. F., et al. "Electrical Guidance of Human Stem Cells in the Rat Brain." *Stem Cell Reports*, vol. 9, no. 1 (2017), pp. 177–89

Chapter 8: At the end

1 Rose, Sylvan Meryl, and H. M. Wallingford. "Transformation of renal tumors of frogs to normal tissues in regenerating limbs of salamanders." *Science*, vol. 107, no. 2784 (1948), p. 457
2 Oviedo, Néstor J., and Wendy S. Beane. "Regeneration: The origin of cancer or a possible cure?" *Seminars in Cell & Developmental Biology*, vol. 20, no. 5 (2009), pp. 557–64
3 Fatima, Iqra, et al. "Skin Aging in Long-Lived Naked Mole-Rats Is Accompanied by Increased Expression of Longevity-Associated and Tumor Suppressor Genes." *Journal of Investigative Dermatology*, 9 June 2022, doi: 10.1016/j.jid.2022.04.028
4 Ruby, J. Graham, et al. "Naked mole-rat mortality rates defy Gompertzian laws by not increasing with age." *eLife* 7:e31157 (2018), doi: 10.7554/eLife.31157
5 Burr, Harold Saxton. *Blueprint for Immortality: The Electric Patterns of Life*. Essex: Neville Spearman Publishers, 1972, p. 53
6 Burr, Harold Saxton. *Blueprint for Immortality: The Electric Patterns of Life*. Essex: Neville Spearman Publishers, 1972, p. 54
7 Langman, Louis, and Burr, H. S. "Electrometric Studies in Women with Malignancy of Cervix Uteri." *Science*, vol. 105, no. 2721 (1947), pp. 209–10
8 Langman, Louis, and Burr, H. S. "A technique to aid in the detection of malignancy of the female genital tract." *Journal of the American Journal of Obstetrics and Gynecology*, vol. 57, no. 2 (1949), pp. 274–81
9 Langman & Burr, "Electrometric," p. 210
10 Stratton, M. R. (2009). "The cancer genome." *Nature*, vol. 458, article 7239 (2009), pp. 719–24, doi: 10.1038/nature07943

NOTES

11 Nordenström, Björn. "Biologically closed electric circuits: Activation of vascular-interstitial closed electric circuits for treatment of inoperable cancers." *Journal of Bioelectricity* 3 (1984), pp. 137–53
12 Nordenström, Björn. *Biologically Closed Electric Circuits: Clinical, Experimental, and Theoretical Evidence for an Additional Circulatory System*. Stockholm: Nordic Medical Publications, 1983
13 Nordenström, *Biologically closed*
14 Nordenström, *Biologically closed*, p. vii
15 Parachini, Allan. "Cancer-Treatment Theory an Enigma to Scientific World," *Los Angeles Times*, 30 September 1986 <https://www.latimes.com/archives/la-xpm-1986-09-30-vw-10015-story.html>
16 Parachini, "Cancer-Treatment," 1986
17 Nordenström, "Biologically closed"
18 Parachini, "Cancer-Treatment," 1986
19 Nilsson E., et al. "Electrochemical treatment of tumours." *Bioelectrochemistry*, vol. 51, no. 1 (2000), pp. 1–11
20 All statistics from "Proceedings of the International Association for Biologically Closed Electric Circuits." *European Journal of Surgery 1994 Supplement* 574, pp. 7–23
21 "Activation of BCEC-channels for Electrochemical Therapy (ECT) of Cancer." *Proceedings of the IABC International Association for Biologically-Closed Electric Circuits (BCEC) in Medicine and Biology*. Stockholm, September 12–15, 1993 (1994), pp. 25–9 <https://pubmed.ncbi.nlm.nih.gov/7531011/>
22 "Björn Nordenström," *20/20*, ABC News, first broadcast 21 October 1988. Available on YouTube: <https://www.youtube.com/watch?v=OmqTKh-CP88>
23 Moss, Ralph W. "Bjorn E. W. Nordenström, MD." *Townsend Letter, The Examiner of Alternative Medicine* 285 (2007), p. 156 <link.gale.com/apps/doc/A162234818/AONE?u=anon~51eea7d2&sid=bookmark-AONE&xid=8719a268>. Accessed 5 August 2021
24 Lois, Carlos, and Arturo Alvarez-Buylla. "Long-distance neuronal migration in the adult mammalian brain." *Science* 264 (1994), pp. 1145–8, doi: 10.1126/science.8178174
25 Grimes, J. A., et al. "Differential expression of voltage-activated Na + currents in two prostatic tumour cell lines: contribution to invasiveness in vitro." *FEBS Letters* 369 (1995), pp. 290–4 <https://febs.onlinelibrary.wiley.com/doi/epdf/10.1016/0014-5793%2895%2900772-2>

26 Reported ubiquitously, including in Pullar, Christine E. (ed.). *The Physiology of Bioelectricity in Development, Tissue Regeneration and Cancer.* Boca Raton: CRC Press, 2011, p. 271

27 Arcangeli, Annarosa, and Andrea Becchetti. "New Trends in Cancer Therapy: Targeting Ion Channels and Transporters." *Pharmaceuticals*, vol. 3, no. 4 (2010), pp. 1202–24

28 Bianchi, Laura, et al. "hERG Encodes a K+ Current Highly Conserved in Tumors of Different Histogenesis: A Selective Advantage for Cancer Cells?" *Cancer Research*, vol. 58, no. 4 (1998), pp. 815–22

29 Kunzelmann, 2005; Fiske, et al., 2006; Stuhmer, et al., 2006; Prevarskaya, et al., 2010; Becchetti, 2011; Brackenbury, 2012, collected in Yang Ming and William Brackenbury. "Membrane potential and cancer progression." *Frontiers in Physiology*, vol. 4, article 185 (2013), doi: https://doi.org/10.3389/fphys.2013.00185

30 Santos, Rita, et al. "A comprehensive map of molecular drug targets." *Nature Reviews Drug Discovery*, vol. 16, no. 1 (2017), pp. 19–34

31 McKie, Robin. "For 30 years I've been obsessed by why children get leukaemia. Now we have an answer," *The Guardian*, 30 December 2018 <https://www.theguardian.com/science/2018/dec/30/children-leukaemia-mel-greaves-microbes-protection-against-disease>

32 Djamgoz, Mustafa, S. P. Fraser, and W. J. Brackenbury. "In Vivo Evidence for Voltage-Gated Sodium Channel Expression in Carcinomas and Potentiation of Metastasis." *Cancers*, vol. 11, no. 11 (2019), p. 1675

33 Leanza, Luigi, Antonella Managò, Mario Zoratti, Erich Gulbins, and Ildiko Szabo. "Pharmacological targeting of ion channels for cancer therapy: In vivo evidences." *Biochimica et Biophysica Acta (BBA)—Molecular Cell Research*, vol. 1863, no. 6, Part B (2016), pp. 1385–97

34 In 2019, a Chinese multicentre preclinical trial tested an antibody that was effective against Djamgoz's variant in mice. They claimed this was able to suppress metastasis. Gao, R., et al. "Nav1.5-E3 antibody inhibits cancer progression." *Translational Cancer Research*, vol. 8, no. 1 (2019), pp. 44-50, doi: 10.21037/tcr.2018.12.23

35 Lang, F., and C. Stournaras. "Ion channels in cancer: Future perspectives and clinical potential." *Philosophical Transactions of the Royal Society of London. Series B, Biological sciences*, vol. 369, article

1638 (2014), 20130108 <https://www.ncbi.nlm.nih.gov/pmc/articles/PMC3917362/pdf/rstb20130108.pdf>

36 "An interview with Professor Mustafa Djamgoz," External Speaker Series presentation, Metrion BioSciences, Cambridge 2018

37 "The Bioelectricity Revolution: A Discussion Among the Founding Associate Editors." *Bioelectricity*, vol. 1, no. 1 (2019), pp. 8–15

38 Greaves, Mel. "Nothing in cancer makes sense except . . . ." *BMC Biology*, vol. 16, no. 22 (2018)

39 Wilson, Clare. "The secret to killing cancer may lie in its deadly power to evolve," *New Scientist*, 4 March 2020 <https://www.newscientist.com/article/mg24532720-800-the-secret-to-killing-cancer-may-lie-in-its-deadly-power-to-evolve/>

40 Hope, Tyna, and Siân Iles. "Technology review: The use of electrical impedance scanning in the detection of breast cancer." *Breast Cancer Research*, vol. 6, no. 69 (2004), pp. 69–74

41 Wilke, Lee, et al. "Repeat surgery after breast conservation for the treatment of stage 0 to II breast carcinoma: a report from the National Cancer Data Base, 2004–2010." *JAMA Surgery*, vol. 149, no. 12 (2014), pp. 1296–305

42 Dixon, J. Michael, et al. "Intra-operative assessment of excised breast tumour margins using ClearEdge imaging device." *European Journal of Surgical Oncology* 42 (2016), pp. 1834–40, doi: 10.1016/j.ejso.2016.07.141

43 Djamgoz, Mustafa. "In vivo evidence for expression of voltage-gated sodium channels in cancer and potentiation of metastasis," Sophion Bioscience YouTube channel, 18 July 2019 <https://www.youtube.com/watch?v=bkKewfmCW6A>. The relevant section of the lecture begins around sixteen minutes in.

44 Dokken, Kaylinn, and Patrick Fairley. "Sodium Channel Blocker Toxicity" [Updated 30 April 2022]. In: StatPearls [Internet]. Treasure Island, FL: StatPearls Publishing, 2022 <https://www.ncbi.nlm.nih.gov/books/NBK534844/>

45 Reddy, Jay P., et al. "Antiepileptic drug use improves overall survival in breast cancer patients with brain metastases in the setting of whole brain radiotherapy." *Radiotherapy and Oncology*, vol. 117, no. 2 (2015), pp. 308–14, doi: 10.1016/j.radonc.2015.10.009

46 Takada, Mitsutaka, et al. "Inverse Association between Sodium

Channel-Blocking Antiepileptic Drug Use and Cancer: Data Mining of Spontaneous Reporting and Claims Databases." *International Journal of Medical Sciences*, vol. 13, no. 1 (2016), pp. 48–59, doi: 10.7150/ijms.13834

47 "An interview with Professor Mustafa Djamgoz," External Speaker Series presentation, Metrion BioSciences, Cambridge 2018

48 Quail, Daniela F., and Johanna A. Joyce. "Microenvironmental regulation of tumor progression and metastasis." *Nature Medicine*, vol. 19, no. 11 (2013), pp. 1423–37, doi: 10.1038/nm.3394

49 Zhu, Kan, et al. "Electric Fields at Breast Cancer and Cancer Cell Collective Galvanotaxis." *Scientific Reports*, vol. 10, no. 1 (2020), article 8712

50 Wapner, Jessica. "A New Theory on Cancer: What We Know About How It Starts Could All Be Wrong," *Newsweek*, 17 July 2017 <https://www.newsweek.com/2017/07/28/cancer-evolution-cells-637632.html>; see also Davies, Paul. "A new theory of cancer," *The Monthly*, November 2018 <https://www.themonthly.com.au/issue/2018/november/1540990800/paul-davies/new-theory-cancer#mtr>

51 Silver, Brian, and Celeste Nelson. "The Bioelectric Code: Reprogramming Cancer and Aging From the Interface of Mechanical and Chemical Microenvironments." *Frontiers in Cell and Developmental Biology*, vol. 6, no. 21 (2018)

52 Lobikin, Maria, Brook Chernet, Daniel Lobo, and Michael Levin. "Resting potential, oncogene-induced tumorigenesis, and metastasis: the bioelectric basis of cancer in vivo." *Physical Biology*, vol. 9, no. 6 (2012), loc. 065002. doi: 10.1088/1478-3975/9/6/065002

53 Chernet, Brook, and Michael Levin. "Endogenous Voltage Potentials and the Microenvironment: Bioelectric Signals that Reveal, Induce and Normalize Cancer." *Journal of Clinical and Experimental Oncology*, Suppl. 1:S1-002 (2013), doi: 10.4172/2324-9110

54 Chernet & Levin, "Endogenous"

55 Gruber, Ben. "Battling cancer with light," Reuters, 26 April 2016 <https://www.reuters.com/article/us-science-cancer-optogenetics-idUSKCN0XN1U9>

56 Chernet, Brook, and Michael Levin. "Transmembrane voltage

potential is an essential cellular parameter for the detection and control of tumor development in a Xenopus model." *Disease Models & Mechanisms*, vol. 6, no. 3 (2013), pp. 595–607, doi: 10.1242/dmm.010835
57 Silver & Nelson, "The Bioelectric Code"
58 Tuszynski, Jack, Tatiana Tilli, and Michael Levin. "Ion Channel and Neurotransmitter Modulators as Electroceutical Approaches to the Control of Cancer." *Current Pharmaceutical Design*, vol. 23, no. 32 (2017), pp. 4827–41
59 Schlegel, Jürgen, et al. "Plasma in cancer treatment." *Clinical Plasma Medicine*, vol. 1, no. 2 (2013), pp. 2–7

Part 5: Bioelectricity in the Future

Chapter 9: Swapping silicon for squids

1 Brown, Joshua. "Team Builds the First Living Robots," The University of Vermont, 13 January 2020 <https://www.uvm.edu/news/story/team-builds-first-living-robots>
2 Lee, Y., et al. "Hydrogel soft robotics." *Materials Today Physics* 15 (2020) <https://doi.org/10.1016/j.mtphys.2020.100258>
3 Thubagere, Anupama, et al. "A Cargo-Sorting DNA Robot." *Science*, vol. 357, article 6356 (2017), eaan6558
4 Solon, Olivia. "Electroceuticals: Swapping drugs for devices," *Wired*, 28 May 2013 <https://www.wired.co.uk/article/electroceuticals>
5 Geddes, Linda. "Healing spark: Hack body electricity to replace drugs," *New Scientist*, 19 February 2014 <https://www.newscientist.com/article/mg22129570-500-healing-spark-hack-body-electricity-to-replace-drugs/>
6 Behar, Michael. "Can the nervous system be hacked?" *The New York Times*, 23 May 2014 <https://www.nytimes.com/2014/05/25/magazine/can-the-nervous-system-be-hacked.html>
7 Mullard, Asher. "Electroceuticals jolt into the clinic, sparking autoimmune opportunities." *Nature Reviews Drug Discovery* 21 (2022), pp. 330–1
8 Hoffman, Henry, and Harold Norman Schnitzlein. "The Numbers of Nerve Fibers in the Vagus Nerve of Man." *The Anatomical Record*, vol. 139, no. 3 (1961), pp. 429–35

9. Davies, Dave. "Are Implanted Medical Devices Creating a 'Danger Within Us'?" NPR, 17 January 2018 <https://www.npr.org/2018/01/17/578562873/are-implanted-medical-devices-creating-a-danger-within-us>
10. Golabchi, Asiyeh, et al. "Zwitterionic Polymer/Polydopamine Coating Reduce Acute Inflammatory Tissue Responses to Neural Implants." *Biomaterials* 225 (2019), 119519 <https://doi.org/10.1016/j.biomaterials.2019.119519>
11. Leber, Moritz, et al. "Advances in Penetrating Multichannel Microelectrodes Based on the Utah Array Platform." In: Xiaoxiang Zheng (ed.), *Neural Interface: Frontiers and Applications*. Singapore: Springer, 2019, pp. 1–40
12. Yin, Pengfei, et al. "Advanced Metallic and Polymeric Coatings for Neural Interfacing: Structures, Properties and Tissue Responses." *Polymers*, vol. 13, no. 16 (2021), article 2834 <https://www.ncbi.nlm.nih.gov/pmc/articles/PMC8401399/pdf/polymers-13-02834.pdf>
13. Aregueta-Robles, U. A., et al. "Organic electrode coatings for next-generation neural interfaces." *Frontiers in Neuroengineering*, 27 May 2014 <https://doi.org/10.3389/fneng.2014.00015>
14. "The Nobel Prize in Chemistry 2000," NobelPrize.org <https://www.nobelprize.org/prizes/chemistry/2000/summary/>
15. Cuthbertson, Anthony. "Material Found by Scientists 'Could Merge AI with Human Brain,'" *The Independent*, 17 August 2020 <https://www.independent.co.uk/tech/artificial-intelligence-brain-computer-cyborg-elon-musk-neuralink-a9673261.html>
16. Chen, Angela. "Why It's so Hard to Develop the Right Material for Brain Implants," *The Verge*, 30 May 2018 <https://www.theverge.com/2018/5/30/17408852/brain-implant-materials-neuroscience-health-chris-bettinger>
17. Technically, there are also ways to inhibit action potentials, but that just means stimulating inhibitory neurons—which are the kinds of neurons that make other neurons not fire. But it's still the same mechanism.
18. Some companies try to understand how the body has interpreted the action potential by implanting even more electrodes to listen to the ensuing signals. But that carries additional surgical risk, and it's certainly not happening in humans.
19. Casella, Alena, et al. "Endogenous Electric Signaling as a Blueprint

for Conductive Materials in Tissue Engineering." *Bioelectricity*, vol. 3, no. 1 (2021), pp. 27–41

20 Demers, Caroline, et al. "Natural Coral Exoskeleton as a Bone Graft Substitute: A Review." *Bio-Medical Materials and Engineering*, vol. 12, no. 1 (2002), pp. 15–35

21 Israel-based OkCoral and CoreBone grow coral on a special diet to make it especially suitable to grafting.

22 Wan, Mei-chen, et al. "Biomaterials from the Sea: Future Building Blocks for Biomedical Applications." *Bioactive Materials*, vol. 6, no. 12 (2021), pp. 4255–85

23 DeCoursey, Thomas. "Voltage-Gated Proton Channels and Other Proton Transfer Pathways." *Physiological Reviews*, vol. 83, no. 2 (2003) pp. 475–579, doi: 10.1152/physrev.00028.2002

24 Lane, Nick. "Why Are Cells Powered by Proton Gradients?." *Nature Education*, vol. 3, no. 9 (2010), p. 18

25 Kautz, Rylan, et al. "Cephalopod-Derived Biopolymers for Ionic and Protonic Transistors." *Advanced Materials*, vol. 30, no. 19 (2018), loc. 1704917

26 Ordinario, David, et al. "Bulk protonic conductivity in a cephalopod structural protein." *Nature Chemistry*, vol. 6, no. 7 (2014), pp. 596–602

27 Strakosas, Xenofon, et al. "Taking Electrons out of Bioelectronics: From Bioprotonic Transistors to Ion Channels." *Advanced Science*, vol. 4, no. 7 (2017), loc. 1600527

28 Kim, Young Jo, et al. "Self-Deployable Current Sources Fabricated from Edible Materials." *Journal of Materials Chemistry B* 31 (2013), p. 3781, doi: 10.1039/C3TB20183J

29 Ordinario, David, et al. "Protochromic Devices from a Cephalopod Structural Protein." *Advanced Optical Materials*, vol. 5, no. 20 (2017), loc. 1600751

30 Sheehan, Paul. "Bioelectronics for Tissue Regeneration." Defense Advanced Projects Research Agency <https://www.darpa.mil/program/bioelectronics-for-tissue-regeneration>. Accessed 31 May 2022

31 Kriegman, Sam, et al., "Kinematic Self-Replication in Reconfigurable Organisms." *Proceedings of the National Academy of Sciences*, vol. 118, no. 49 (2021), loc. e2112672118 <https://doi.org/10.1073/pnas.2112672118>

32 Coghlan, Simon, and Kobi Leins. "Will self-replicating 'xenobots' cure diseases, yield new bioweapons, or simply turn the whole world into grey goo?" The Conversation, 9 December 2021 <https://theconversation.com/will-self-replicating-xenobots-cure-diseases-yield-new-bioweapons-or-simply-turn-the-whole-world-into-grey-goo-173244>
33 Adamatzky, Andrew, et al. "Fungal Electronics." *Biosystems* 212 (2021), loc. 104588, doi: 10.1016/j.biosystems.2021.104588

Chapter 10: Electrifying ourselves better

1 Nitsche, Michael A., et al. "Facilitation of Implicit Motor Learning by Weak Transcranial Direct Current Stimulation of the Primary Motor Cortex in the Human." *Journal of Cognitive Neuroscience*, vol. 15, no. 4 (2003), pp. 619–26, doi: https://doi.org/10.1162/089892903321662994
2 Trivedi, Bijal. "Electrify your mind—literally," *New Scientist*, 11 April 2006 < https://www.newscientist.com/article/mg19025471-100-electrify-your-mind-literally/>
3 Marshall, L., M. Mölle, M. Hallschmid, and J. Born. "Transcranial direct current stimulation during sleep improves declarative memory." *The Journal of Neuroscience*, vol. 24, no. 44 (2004), pp. 9985–92, doi: 10.1523/Jneurosci.2725-04.2004
4 Walsh, Professor Vincent. "Cognitive Effects of TDC at Summit on Transcranial Direct Current Stimulation (tDCS) at the UC-Davis Center for Mind & Brain," UC Davis YouTube channel, 8 October 2013 <https://www.youtube.com/watch?v=9fz7r8VDV4o>. The relevant section of the lecture begins around fourteen minutes in.
5 Wurzman, Rachel, et al. "An open letter concerning do-it-yourself users of transcranial direct current stimulation." *Annals of Neurology*, vol. 80, no. 1. July 2016
6 Nord, Camilla, et al. "Neural predictors of treatment response to brain stimulation and psychological therapy in depression: a double-blind randomized controlled trial," *Neuropsychopharmacology*, vol 44, no. 9 (2019), pp. 1613–1622
7 Aschwanden, Christie. "Science isn't broken: It's just a hell of a lot harder than we give it credit for," Five Thirty-Eight, 19 August 2015 <https://fivethirtyeight.com/features/science-isnt-broken/>

NOTES

8   Verma, N., et al. "Auricular Vagus Neuromodulation—A Systematic Review on Quality of Evidence and Clinical Effects." *Frontiers in Neuroscience* 15 (2021), article 664740 <https://doi.org/10.3389/fnins.2021.664740>
9   Young, Stella. "I'm not your inspiration, thank you very much." TED, June 2014, www.ted.com/talks/stella_young_i_m_not_your_inspiration_thank_you_very_much/
10 Sierra-Mercado, Zuk, Beauchamp, et al. "Device Removal Following Brain Implant Research." Neuron 103 (2019), pp. 759–761 <https://doi:10.1016/j.neuron.2019.08.24>
11 Source is interview with the author at the International Neuroethics Society meeting, 2 November 2018. The issues are also explored in Drew, Liam. "The ethics of brain–computer interfaces." *Nature*. 24 July 2019 <https://www.nature.com/articles/d41586-019-02214-2>
12 Strickland, Eliza. "Worldwide Campaign For Neurorights Notches Its First Win," *IEEE Spectrum*, 18 December 2021 <https://spectrum.ieee.org/neurotech-neurorights>
13 Coghlan, Andy. "Vaping really isn't as harmful for your cells as smoking," *New Scientist*, 4 January 2016 <https://www.newscientist.com/article/dn28723-vaping-really-isnt-as-harmful-for-your-cells-as-smoking/>
14 "Committee on the Review of the Health Effects of Electronic Nicotine Delivery Systems and Others." In: Kathleen Stratton, Leslie Y. Kwan, and David L. Eaton (eds), *Public Health Consequences of E-Cigarettes*, Washington, DC: 2018, 24952 <https://www.nap.edu/catalog/24952>
15 Ozekin, Yunus, Kayla Moehn, Emily Bates, et al. "Intrauterine Exposure to Nicotine Through Maternal Vaping Disrupts Embryonic Lung and Skeletal Development via the Kcnj2 Potassium Channel." *Developmental Biology*, vol. 501 (2023), pp. 111–23 <https://doi.org/10.1016/j.ydbio.2023.06.002>
16 Benzonana, Laura, et al. "Isoflurane, a Commonly Used Volatile Anesthetic, Enhances Renal Cancer Growth and Malignant Potential via the Hypoxia-Inducible Factor Cellular Signaling Pathway In Vitro." *Anesthesiology*, vol. 119, no. 3 (2013), pp. 593–605
17 Jiang, Jue, and Hong Jiang. "Effect of the Inhaled Anesthetics

Isoflurane, Sevoflurane and Desflurane on the Neuropathogenesis of Alzheimer's Disease (Review)." *Molecular Medicine Reports*, vol. 12, no. 1 (2015), pp. 3–12

18 Robson, David. "This is what it's like waking up during surgery," Mosaic, 12 March 2019 <https://mosaicscience.com/story/anaesthesia-anesthesia-awake-awareness-surgery-operation-or-paralysed/>

19 Edelman, Elazer, et al. "Case 30-2020: A 54-Year-Old Man with Sudden Cardiac Arrest." *New England Journal of Medicine*, vol. 383, no. 13 (2020), pp. 1263–75

20 Hesham, R. Omar, et al. "Licorice Abuse: Time to Send a Warning Message." *Therapeutic Advances in Endocrinology and Metabolism*, vol. 3, no. 4 (2012), pp. 125–38

21 Actually, I noticed two patterns: most of the scientists who got hit with the most scathing criticism were women. The men sometimes didn't recall any trouble at all.

22 Davies, Paul. *The Demon in the Machine*. London: Allen Lane, 2019, p. 86

23 McNamara, H. M., et al. "Bioelectrical domain walls in homogeneous tissues." *Nature Physics* 16 (2020), pp. 357–64 <https://doi.org/10.1038/s41567-019-0765-4>

24 Davies, *The Demon in the Machine*, pp. 82–3

25 Pietak, A., and Levin, M. "Exploring Instructive Physiological Signaling with the Bioelectric Tissue Simulation Engine." *Frontiers in Bioengineering and Biotechnology*, vol. 4, article 55 (2016), doi: 10.3389/fbioe.2016.00055

26 Hutchinson, Alex. "Is Brain Stimulation the Next Big Thing?" *Outside*, October 24, 2019, <https://www.outsideonline.com/health/training-performance/neurofire-brain-stimulation-tdcs-bike-tour/>

Afterword: Gut feelings

1 Alvarez, Walter. "The Electrogastrogram and What It Shows." *Journal of the American Medical Association* vol. 78, no. 15 (1922), pp. 1116–119, doi:10.1001/jama.1922.02640680020008

2 Blair, Peter J., et al. "The Significance of Interstitial Cells in Neurogastroenterology." *Journal of Neurogastroenterology and Motility* vol. 20, no. 3 (2014), pp. 294–317 <https://doi.org/10.5056/jnm14060>

NOTES

3   Baker, Salah A., et al. "Ca2+ signaling driving pacemaker activity in submucosal interstitial cells of Cajal in the murine colon." *eLife* (2021), 10:e64099

4   Feloney, M. P., K. Stauss, and S. W. Leslie. *Sacral Neuromodulation*. [Updated May 30, 2023]. In: StatPearls. Treasure Island, FL: StatPearls Publishing, 2022 <https://www.ncbi.nlm.nih.gov/books/NBK567751/>

5   Reddit. "CNN article on Wegovy/Ozempic." <https://www.reddit.com/r/MaintenancePhase/comments/159dgvn/cnn_article_on_wegovyozempic/>

6   Zoll, B., et al. "Gastric Electric Stimulation for Refractory Gastroparesis." *Journal of Clinical Outcomes Management* vol. 26, no. 1 (Jan. 2019), pp. 27–38 <https://www.ncbi.nlm.nih.gov/pmc/articles/PMC6733037/>

7   Kallmes, D. F., and R. M. Ruedy. "Humanitarians, compassion, and the Food and Drug Administration: guidance for the practitioner." *American Journal of Neuroradiology* vol. 30, no. 2 (Feb. 2009), pp. 216–18, doi: 10.3174/ajnr.A1373

8   US Food and Drug Administration. "Humanitarian Device Exemption." <https://www.fda.gov/medical-devices/premarket-submissions-selecting-and-preparing-correct-submission/humanitarian-device-exemption>

9   Gharibans, A. A., et al. "Gastric dysfunction in patients with chronic nausea and vomiting syndromes defined by a noninvasive gastric mapping device." *Science Translational Medicine* vol. 14, no. 663 (Sept. 21, 2022), eabq3544, doi: 10.1126/scitranslmed.abq3544

10 Ignacio, Rebollo, et al. "Stomach-brain synchrony reveals a novel, delayed-connectivity resting-state network in humans." *eLife* 7:e33321 (2018).

11 Choe, A. S., et al. "Phase-locking of resting-state brain networks with the gastric basal electrical rhythm." *PLOS One*, January 5, 2021 <https://doi.org/10.1371/journal.pone.0244756>

12 Sagar, Soumya. "This microbe-filled pill could track inflammation in the gut." *MIT Technology Review*, October 18, 2023 <https://www.technologyreview.com/2023/10/18/1081842/this-microbe-filled-pill-could-track-inflammation-in-the-gut/>

13 Nord, Camilla. *The Balanced Brain*. London: Allen Lane, 2023, p. 43

14 Nord, C., et al. "A causal role for gastric rhythm in human disgust avoidance," *Current Biology* vol. 31 (Feb. 8, 2021), pp. 629 and 634
15 Lincoff, A. Michael, et al. "Semaglutide and cardiovascular outcomes in obesity without diabetes," *New England Journal of Medicine*. November 11, 2023, DOI: 10.1056/NEJMoa2307563
16 The Cleveland Clinic. "GLP-1 agonists: what they are, how they work, and side effects." *Treatments and Procedures Fact Sheet*. [retrieved https://my.clevelandclinic.org/health/treatments/13901-glp-1-agonists]
17 Curley, Mike. "Eli Lilly and Novo Nordisk sued over risks of diabetes drugs," *Law 360*. August 2, 2023 <https://www.law360.com/articles/1706667/eli-lilly-and-novo-nordisk-sued-over-risks-of-diabetes-drugs>
18 Jones, Diana Novak. "As Ozempic cases mount, consumer lawyers push to consolidate lawsuits." *Reuters*, December 4, 2023. <https://www.reuters.com/legal/legalindustry/ozempic-cases-mount-consumer-lawyers-push-consolidate-lawsuits-2023-12-04/>
19 Jones, Diana Novak. "Ozempic side effects 'well-known,' Novo Nordisk argues," *Reuters*, November 6, 2023 <https://www.reuters.com/legal/litigation/ozempic-side-effects-well-known-novo-nordisk-argues-2023-11-06/>, See also <https://fingfx.thomsonreuters.com/gfx/legaldocs/zgpordwdwvd/novoMTD.pdf>
20 Huang, I. H., et al. "Worldwide prevalence and burden of gastroparesis-like symptoms as defined by the United European Gastroenterology (UEG) and European Society for Neurogastroenterology and Motility (ESNM) consensus on gastroparesis." *United European Gastroenterol Journal* vol. 10, no. 8 (Oct. 2022), pp. 888–97, doi: 10.1002/ueg2.12289
21 Zajdel, Tom, et al. "SCHEEPDOG: programming electric cues to dynamically herd large-scale cell migration." *Cell Systems*, vol. 10, no. 6 (June 24, 2022), P506–514.E3
22 Adee, Sally. "The farmers boosting crops with electricity," *BBC Future*. August 17, 2023 <https://www.bbc.com/future/article/20230816-the-farmers-boosting-crops-with-electricity>

INDEX

Abell, Tom 296–7, 306, 309
Académie des sciences (France) 37, 40, 50
action potential 74–6, 78–9, 80–1, 86–8
  and cancer 224–5, 226, 227
  and the heart 90, 104, 108
  and implants 253
  and the spine 173
Adamatzky, Andrew 265
Adams, Dany Spencer 201–2, 203, 204, 205–6, 287, 291
  and cancer treatments 232–3, 238, 239
  and protons 256, 257
Adrian, Edgar 113, 114–17, 122, 123–4, 135
  and Berger 119, 121
afferent system 74
aggression 126
Agnew, William 83
AI (artificial intelligence) 292
alcohol 285
Aldini, Giovanni 29, 39, 47–9, 105, 106
  and Galvani 49–53
  and resuscitation 53–6
algae 91, 152–4
alkaline earth metals 60
Allen, Paul 211
alpha waves 136
Altmann, Margaret 192
Alvarez, Walter 297
Alzheimer's disease 136
amber 22, 23, 24
Ampère, André-Marie 63
amputation 187

anaesthesia 285
Ancient Greece 22
Andara 165, 168, 170, 172, 173
animal electricity 47, 48–53, 59–60, 61–3, 67
  and the brain 125–7
  and ovulation 191, 192
  and regeneration 206–7, 209
  and the spine 157, 158
  see also dogs; frogs
animal spirits (*pneuma psychikon*) 20, 21, 28
anti-vivisectionism 99–100
Arcangeli, Annarosa 226, 228, 235
Arrhenius, Svante 77
arthropods 258
artificial electricity *see* batteries
Ashcroft, Frances 89, 90
  The Spark of Life 85
atria 103
autism 131, 274
Aw, Sherry 205–6
axons 74–5, 78, 87–8, 157–9, 173

bacteria 10, 19, 92, 95
Badylak, Stephen 188–9, 211, 288
Bassi, Laura 28, 33
Bates, Emily 283–4, 285, 287–8, 289–90
batteries 43, 44, 45, 46, 60
  and Aldini 52–3
  and frogs 63–5
  and pacemakers 107–8
Beccaria, Giambattista 28, 33–4
bed sores 180
Benabid, Alim-Louis 129–30
Benedict XIV, Pope 26

Berger, Hans   11, 117, 118–22, 271
Berger, Theodore   110–11, 141–2, 143–4
beta waves   136
BETR programme   260–1
BETSE (Bioelectric Tissue Simulation Engine)   292
Bettinger, Chris   250
biocompatible materials   244–5
bioelectric code   94–6, 205–6, 210
Bioelectricity (journal)   290
bioimpedance   231–3
biology   19, 21, 60
bioreactor   187
Bird, Golding   60–1
birth control   192, 193
birth defects   10, 283–5
Bissel, Mina   201
Blondel, Christine   37, 53
blood   9, 19, 103–4, 105
Bohnert, Debra   160, 161, 163, 166, 171, 172
Bologna *see* University of Bologna
bones   9, 27, 179, 202, 258–9
  and coral grafts   254
  and healing   180
Bongard, Joshua   243–4, 264
Borelli, Alfonso   20
Borgens, Richard   155–9, 160–3, 165–6, 167, 168, 171–4, 183
Bouton, Chad   139–40
bradycardia   105, 108
brain, the   9, 11, 13, 52–3, 117–22
  and chips   144–9
  and computing   122–3
  and ECG   109
  and electro-therapy   123–5
  and implants   110–11, 125–7, 128–34, 248–51, 252–3
  and neurons   74
  and tumours   118, 119, 136
  see also DBS; memory; neural code; tDCS
BrainGate   138–9, 145–6
Bresadola, Marco   26, 46, 69

Brugnatelli, Valentino   39, 41, 45
Buoniconti, Marc   159–60
Burkhardt, Ian   139–40, 144–5, 149
Burr, Harold Saxton   189–94, 195, 198, 215–17, 231, 293
Byron, Lord   59

calcium   8, 76, 79, 92
  and cancer   228
  and channels   83, 87–8, 90
  and sperm   195, 196
Campenot, Robert   86
cancer   10, 11–12, 13, 235–9
  and ion channels   91, 216–22, 224–9, 239–40
  and regeneration   214–15
  and treatment   229–30, 231–5, 261–2
Carpue, Joseph   48, 53
Carradori, Giovacchino   39, 41, 45
Catholicism   26, 27, 33, 193
Caton, Richard   117–18
Cavuoto, James   151, 165, 170–1
Celestial Bed   57
cell membrane   82, 84, 86–7, 94–6
cephalopods *see* squid
Chernet, Brook   238–9
children   207–8
chitosan   255, 257
chloride   76, 77, 79, 83
  and cancer   228
  and ovulation   194
  and sperm   195
cilia   263
ClearEdge   232
cloning   83
Cobb, Matthew   123
coding   13–14, 122–3, 142
  and bioelectric   94–6, 205–6, 210
  and neural   111–17, 134–41
Cohen, Adam   291
collagen   255
computers   122–3
Connecticut Medical Society   57–8
consciousness   134, 137
Copeland, Nathan   141

INDEX

Copernicus, Nicolaus 19
coral 254
Cormie, Peter 167
corpses 47–9, 54–5
Coulomb, Charles 37
Covid-19 pandemic 168, 172, 228, 276–7
Crick, Francis 81, 83, 137, 217
  The Astonishing Hypothesis: The Scientific Search for the Soul 134–5
CRISPR 13–14
Curt, Gregory 219
Cyberkinetics 165–6, 167–8, 170–1
cybernetics 123, 294

DARPA (Defense Advanced Research Project Agency) 142, 182, 259, 260–1
  and regeneration 210–11
  and tDCS 266–70
Davies, Paul 93, 291–2
DBS ('deep brain stimulation') 3, 129–34, 147, 248, 281
De Loof , Arnold 292
death 53–4, 55–6; *see also* corpses
defibrilation 53–4
Delgado, José 125–7
dementia 119, 136, 141–2
dendrites 74–5
depression 3, 5, 7, 52–3, 136, 304
  and DBS 131–3
  and tDCS 276
Dermacorder 181, 182
Descartes, René 20
diabetes 3–4, 89, 90, 304
die-back 157–8
Dixon, Mike 232
Djamgoz, Mustafa 11–12, 14, 222–9, 233–5
DNA 81, 83, 94–5, 217
Dobbs, David 132
dogs 99–100, 160–2
Donoghue, John 138, 146, 170
drugs *see* medicine
Du Bois-Reymond, Emil 65–7, 71, 74, 75

eating disorders 131
ECG (electrocardiogram) 102–3, 109, 148, 189
ECoG (electrocorticography) 136–8, 147
EEG (electroencephalogram) 11, 118–22, 135–6, 271
efferent system 74
EGG (electrogastrogram) 298
eggs 195–6
Einthoven galvanometer 297
Einthoven, Willem 101–3, 104–5
electric fields 92–3
electric fish 22, 25, 28, 62
electrical guidance systems 309
electrical stimulation on gut 298–302
electricity 22–5, 46, 76
  and algae 152–4
  and the brain 117–22
  and embryos 202–4
  and Galvani 18–19, 28–32, 40–1, 42–4
  and the heart 100–9
  and medical care 56–9, 60–1
  and regeneration 209–12
  and resuscitation 53–4, 55–6
  and skin 176–7, 181
  and sperm 195–6
  and the spine 173
  and Volta 17–18, 33–40, 41–2
  and wound-healing 182–3, 260–1
  see also animal electricity; ions; voltage readings
electro-therapy 123–5
electrocardiography 102–3
electroceuticals 3, 245–55, 277
electrogastrography 301
electromagnetism 46
electrome 12–14, 71–2, 91
electrometers 62–3
electrophorus 34–6, 38
electrophysiology 42–3, 72–3, 223–4
electrostatic generators 23, 24, 29
Eli Lilly 307
Ellsworth, Oliver 58

355

embryos 10, 196, 197–206, 291
    and regeneration 207
    and stem cells 188
    see also birth defects
EMG (electromyograph) 189
endothelium 178
Enterra device 301, 307
epigenetics 93
epilepsy 3, 52, 119, 121, 124
    and drugs 234, 283
epithelium 176–9
Essai théorique et expérimental sur le galvanism (Aldini) 51
eyes 178–9, 208, 223–4

Famm, Kris 247
Faraday, Michael 46, 76–7
fat cells 202, 258–9
FDA (Food and Drug Administration) 130, 132, 145, 146–7, 286
    and spinal injury 162, 164, 165–6, 168, 169–71
fertility 192–3, 194
Finger, Stanley 61
Firmian, Carlo 35
First World War 113–14, 116, 124
Forbes, Alexander 114
Forster, George 48–9, 54, 56
Franklin, Benjamin 24, 25, 28, 33
Franklin, Rosalind 81
French, Jennifer 168–9, 278–80, 282
Frisi, Paolo 34
frogs 29–32, 36, 37–41, 42–3
    and du Bois-Reymond 65–6
    and embryos 201–4, 205–6
    and Matteucci 63–5
    and neural code 112, 113, 114–15
    and proton pumps 256, 257
    and regeneration 207, 210–11
    and robotics 243–4, 262–5
    and spinal neurons 154, 166–7
    and tumours 214, 238–9
Frohlich, Flavio 148
fucus 152–4
fungi 10, 92

Galen, Claudius 19–20
Galileo Galilei 19
Galvani, Luigi 11, 18–19, 21, 25–32, 36–46, 59–60, 293
    and Aldini 49–52
    and Humboldt 61
Galvanic Society 51, 52, 53, 55
galvanism 52–3, 59
galvanometers 63, 65, 66; *see also* string galvanometers
gamma waves 136
gap junctions 90–1, 104, 204–5
gastroparesis 296, 299, 301, 307–8
Geisler, David 162
gelatin 255
genome 12, 72, 93–4, 226, 230
George, Mari Hulman 162, 165
Gilbert, Frederic 280
Gilbert, William 22
Gladstone, Herbert 100
GlaxoSmithKline (GSK) 246–7
glia 249–50, 251, 253
Glickman, Morton 220, 221
GLP-1 drugs 306–7
Golgi, Camillo 73
Gomez, Marcella 259–60, 261
Gorodetsky, Alon 257–8
Graham, James 57
Greatbatch, Wilson 107, 108
Greaves, Mel 227, 229–30
Greely, Hank 281–2
Guericke, Otto von 22–3
gut-brain axis, 303–5
gut health 297–8

Hall, Freda 296, 300–1
Hansen, Scott 92
Harari, Yuval Noah: *Homo Deus* 4
Harold, Franklin: *To Make the World Intelligible* 290
Hattersley, Andrew 89, 90
Hauton, Jacques 219
healing 179–83, 211, 215, 260–1
heart 90, 91, 100–5, 111
    and pacemakers 105–9

INDEX

Heath, Robert 125
Heeger, Alan J. 251
Helmholtz, Hermann von 66, 67, 71, 75
hERG channel 226, 228
Hinkle, Laura 156–7, 287
Hodgkin, Alan 75–6, 78–82, 85
homosexuality 125
Humboldt, Alexander von 21, 61–2, 64–5, 66, 67, 71
Hutchinson, Alex 294
Huxley, Andrew 75–6, 78–82, 85
hydras 206–7
hydrogel 244
hydrogen 60, 200
Hyman, Albert 106, 107

Implanted Neural Prosthesis 168–9
implants 245–55, 280–2
Inda, Maria 304
Ingram, Brandon 150–1, 162, 163, 164
interstitial cells of Cajal (ICCs) 298–300, 302–3, 305
ions 8–9, 76–82, 92
 and cancer 218–22, 224–9, 233–5, 239–40
 and channels 82–91, 94–6
 and chitosan 257
 and drugs 282–5
 and embryos 199–200
 and implants 253–4
irritable bowel syndrome (IBS) 308
Jackson, Andrew 244
Jaffe, Lionel 152–5, 156, 159, 183, 195, 198
 and regeneration 209
Johnson, Bryan 110–11, 142–4, 148

Kennedy, Phil 146–7, 279
Kepler, Johannes 19
keratin 255
Kernel 110–11, 142–4, 148, 149
Khalsa, Sahib 304, 309
Kinneir, David 21
Koch, Christoph 137

Langman, Louis 215–17, 231
Langston, Joseph 127–8
Lanzarini, Luigi 52–3
left–right assymetry 198–9, 201, 204
Lenzer, Jeanne: *The Danger Within Us* 248
Levin, Michael 94, 187, 189, 205–6, 292
 and anaesthesia 285
 and cancer treatment 237–8, 239
 and embryos 198, 199–201, 204
 and protons 256
 and regeneration 209–12
 and reviewers 288, 289, 290
 and stem cells 258–60
 and xenobots 262–3, 264
Lewis, Thomas 103
Leyden jars 23–4, 28, 29, 34, 56–7
Li, Jianming 171, 172
lightning 22, 24, 31, 33
liquorice 286
liver 208
Lobikin, Maria 238
Loomis, Alfred 271
Lopez, Jose 240, 288
Lucas, Keith 112, 113–14
Ludwig, Kip 133, 252–3, 274, 276
Lund, Elmer 91

Mayo Clinic, Florida 296
McCaig, Colin 174–6
McCulloch, Warren 122
MacDiarmid, Alan G. 251
MacKinnon, Roderick 83–4
Marblestone, Adam 143–4, 148
margin probes 232–3
marine organisms 254–5
Marshall, John 58
Marshall, Lisa 272
Matteucci, Carlo 63–5
Mayberg, Helen 131, 132, 144, 281
medicine 3–4, 7, 56–9, 60–1
 and cancer 261–2
 and ion channels 89, 90, 226–7, 282–5
 and sodium-channel-blockers 234–5
 see also regenerative medicine

357

medics 21–2
Medtronic 108–9, 129–30, 131–2, 165, 296, 299, 301
MEG (magnetoencephalography) 148
membrane potential 79, 81, 200–2
membrane voltage 200–2, 204–6, 209, 220, 256, 257
  and cancer 232, 237–8
  and stem cells 258–60
memory 110–11, 124, 141–5, 272
Mesmer, Franz 59
Messerli, Mark 180
metals 37, 38, 40–2, 57–8
  and implants 244, 248–50
microscopes 19, 20–1
MIMO (multiple-input/multiple-output) algorithm 141–4
Mitchell, Peter 11
mixed conduction 254
mole rats 215
Morandi, Anna 28
Mounjaro 306, 308
MPTP 128
Müller, Johannes 62, 65
Murder Act (1803) 47–8
Murray, Thomas 294
muscle contraction 29, 30–2, 36, 37–41, 42–3
  and corpses 48–9
  and du Bois-Reymond 66–7
  and neural code 111–16
Musk, Elon 143, 147

Nagle, Matt 138, 277
Napoleon Bonaparte 43–4, 50
natural philosophy 21, 33
Neher, Erwin 82, 83, 90
nerve impulse 74–5, 80–1, 87–8, 111–16, 150–1
nervous system 8–9, 19–21, 71–2, 92–3
  and du Bois-Reymond 66–7
  and electroceuticals 245–7
  and ions 76–81, 91
  and neurons 73–6

neural bypass 138–40
neural code 111–17, 134–41, 142–4
neural dust 147–8
neural lace 147–8
neurograins 147–8
neurons 8–9, 73–6, 249–50, 252–3, 285
Neuropixels 147
neurostimulation 2, 3, 4–7
neurotechnology 278–80
neurotoxins 89–90
neurotransmitters 75, 87–8
Newton, Isaac 19, 21
Nicolelis, Miguel 170
nicotine 284–5
Nitsche, Michael 272
Nobili, Leopoldo 62–3
Noda, Masaharu 83
Nogi, Taisaku 204
Nord, Camilla 304–5, 308–9
Nordenström, Björn 218–21, 239–40
Novo Nordisk 307
Nuccitelli, Richard 12, 153, 155, 156, 173
  and skin 178, 181, 182
Nüsslein-Volhard, Christiane 212

obsessive-compulsive disorder 3, 131, 132, 133
OFS (oscillating field stimulator) 150–2, 159, 160–8, 169–74
O'Grady, Greg 298, 301–4, 309
organic electronics 251–2
organs 9, 178, 212
ovulation 191–4, 215–16
oxygen 60
Ozempic 306, 308

pacemakers 105–9, 128–30, 248
pancreas 89
paralysis *see* spinal injury
paralyzed stomach 296, 301
Parkinson's disease 3, 127–8, 129–30, 133, 136
Paulus, Walter 272
PEDOT 251–2
Penfield, Wilder 124

INDEX

Penninger, Josef 181
Perkins, Elisha 57–9
philosophy 19–20
physics 19, 60
physiological currents 153–7, 174–9
Piccolino, Marco 46, 69
Pietak, Alexis 292
placebo effect 6, 132, 164, 167, 269
planarians 204, 206–7, 210, 285
plants 10, 91–2
plastics 251–2, 254
polyacetylene 251
polymers 244, 251–2, 254
Poo, Mu-ming 154, 155, 156, 158, 177
post-traumatic stress disorder (PTSD) 305
Potamian, Brother 32
potassium 60, 152, 286
   and action potential 77, 79–80, 81
   and cancer 228
   and channels 83–8, 89, 90, 95
   and embryos 199, 200, 205–6
   and nicotine 284–5
   and sperm 195
pregnancy 283–4
prosthetics 13
protein 83, 84
protists 10, 92
protons 199–200, 201, 256–8, 259–60
pseudo-gastroparesis 307
pseudoscience 59
Pullar, Christine 182
Purdon, Patrick 285

quackery 56–9, 121

Rajnicek, Ann 171, 174–6, 177, 181, 287
Ramón y Cajal, Santiago 73
rats 141–2
Ray, Johnny 146
Rebollo, Ignacio 303
reflectin 257–8
regeneration 157–9, 160, 206–11, 256–7
   and cancer 214–15, 261–2

regenerative medicine 187–9, 211–13, 262–3
religion 26–7, 29, 33, 193
repolarization 87–8
resting potential 79–80, 85–6
resuscitation 53–4, 55–6, 106
reviewers 287–90
rheumatoid arthritis 245
Ridgway, Andy 131
Ringer, Sydney 77, 78
Ritter, Johann Wilhelm 61
RNA 94–5
Robinson, Ken 153, 154–5, 156–7, 159, 166–7, 177
   and embryos 199–200
robotics 243–4, 262–5
Rock, John 192, 193–4
Rolandi, Marco 255–6, 257, 258, 259–60, 261
Rose, Sylvan Meryl 214
Royal College of Surgeons (London) 48
Royal Humane Society (London) 55–6
Royal Society (London) 44, 58, 99, 119

Sakmann, Bert 82, 83, 90
salamanders 207, 214
Scheuermann, Jan 146, 149
scientific revolution 22–3
sea stars 207
second brain 302–3
Second World War 123
sensorimotor cortex 140–1
Serafin, Catharina 105, 277
Shapiro, Scott 161, 162–3, 164, 166–8, 171
Sheehan, Paul 259, 260–1, 262
Shelley, Mary: *Frankenstein* 49
Shirakawa, Hideki 251
shocks *see* electro-therapy
sinus node 103–4, 108
Sisken, Betty 209
skin 9, 176–8, 180, 181–2, 209, 258
sleep 121, 135–6, 271
smoking 284
Société philomatique de Paris 40

sodium 60, 152, 210
   and action potential 77, 79–80, 81
   and cancer 225, 227, 228, 233–5
   and channels 83–8, 90, 95
Sonnenschein, Carlos 235–6
Soto, Ana 235–6
Soups 75
Spallanzani, Lazzaro 29–30, 39, 41, 42, 44, 45
   and Aldini 50–1
Sparks 75
sperm 195–6
spinal injury 150–2, 154, 157–74, 278–9
sport 5, 294
squid 78–9, 255–6, 257–8
static electricity 23
stem cells 188–9, 202, 212–13, 208, 258–60
string galvanometers 101–3, 104–5, 114, 118–19
Sundelacruz, Sarah 213
surgery 231–2
Swanton, Francis 229
synapse 75, 104
synaptic plasticity 223–4

tDCS (transcranial direct current stimulation) 2, 3, 4–7, 266–71, 272–6, 293–4
teeth 9
telegraphs 46, 60, 63
telepathy 118
tetrodotoxin 89–90
torpillage 124
Tosti, Elisabetta 195, 199
Tourette's syndrome 131
Tracey, Kevin 245–6
tractors 58–9
transistors 255–6, 257–8
trial volunteers 279–82
Tseng, Ai-Sun 208, 209–10, 287

University of Bologna 27–8, 43–4, 51
Utah array 136–8, 145–6, 147

vagus nerve 245, 246, 247–8, 277
Valli, Eusebio 37, 40, 41
Van Musschenbroek, Pieter 23–4, 33
Vassalli, Anton Maria 41
ventricles 103
Venturoli, Giuseppe 44
Veratti, Giuseppe 28
Volta, Alessandro 17–18, 19, 32–6, 46, 50
   and Faraday 76
   and Galvani 36–45, 51
   and Humboldt 61
voltage readings 189–94, 215–17, 232–3; *see also* membrane voltage

Waller, Augustus 100–1, 102–3, 104, 109, 117, 298
Walsh, John 25, 28, 62
Walsh, Vincent 273–5
Walters, Barbara 221
Washington, George 58
Watson, James 81, 83, 217
Wegovy 306, 308
weight-loss drugs 306
Weisend, Mike 266, 268, 269, 270
Wiener, Norbert: *Cybernetics: Or Control and Communications in the Animal and the Machine* 123
women 28, 189, 190–4, 216–17
wounds 9, 157, 172–3, 177, 178–83, 215
   and chitosan 257
   and types 260–1

xenobots 243–4, 262–5

yeast 19
Young, Stella 278

Zakon, Harold 95, 96
Zepbound 306
Zhao, Min 174–6, 181, 182, 183, 213
Ziemssen, Hugo von 105
Zotterman, Yngve 115–16